TRAVELING THE 38th PARALLEL

A Water Line

Around the World

D0861807

 David and Janet Carle were state park rangers at the Mono Lake Tufa State Reserve for over twenty years, and taught at Cerro Coso Community College in Mammoth Lakes. Janet was editor of the California State Park Rangers Association journal, *The Wave.* David is also the author of:

Mono Lake Viewpoint, Artemisia Press, 1992

Drowning the Dream, California's Water Choices at the Millennium Praeger, 2000

Water and the California Dream, Sierra Club Books, 2003

Burning Questions, America's Fight with Nature's Fire, Praeger, 2002

Introduction to Water in California University of California Press, 2016, 2nd edition

Introduction to Air in California, University of California Press, 2006

Introduction to Fire in California, University of California Press, 2008

Introduction to Earth, Soil, and Land in California University of California Press, 2010

Putting California on the Map: von Schmidt's Lines Phalarope Press, 2018

Bodie's Boss Lawman, The Frontier Odyssey of Constable John F. Kirgan, by Bill Merrell with David Carle, Nevada Publications, 2005

Mono Lake Basin, by David Carle and Don Banta Arcadia Publishing, 2008

Mono, A Novel, Phalarope Press, 2010

My Visit to Mono Lake, Phalarope Press, 2011

ISBN 978-1-0878-7480-7

Cover and Khargilik bazaar photos by Rick Kattelmann
Drake's Bay kayak photo by Ryan Carle
All other photos by David and Janet Carle
Maps prepared by Javier Grijalbo

Phalarope Press

carle@qnet.com

PO Box 39

Lee Vining, California 93541

Printed in the United States of America

First paperback edition, 2020
(Hardcover edition was by University of California Press, 2013)

TRAVELING THE 38ᵗʰ PARALLEL

A Water Line

Around the World

David and Janet Co

For Nick and Ryan, our inspiration

CONTENTS

INTRODUCTION: Parallel Universe 38° North

"You are undertaking the first experience, not of the place, but of yourself in that place. ...for nobody can discover the world for anybody else. It is only after we have discovered it for ourselves that it becomes a common ground and a common bond, and we cease to be alone."
Wendell Berry (1991, 43)

Exactly sixty years after the June 25, 1950 start of the Korean War, we stood at the edge of the Demilitarized Zone (DMZ), a 150-mile long and 2.5-mile wide strip of land on the 38th parallel that spans the Korean Peninsula. A soldier manned a nearby guard tower and stared across the water, ready to strike at any threat from the north. An egret at the river stared into the water, ready to strike at fish. With people excluded from the DMZ for so many decades, the wetlands and forests behind barbed wire have become a haven for wildlife and an inadvertent experiment in ecological recovery.

The Korean experience was part of an around-the-world exploration along the 38th parallel seeking water-related environmental and cultural intersections. Latitude lines are part of an abstract grid that humans have superimposed over the globe. Though imaginary, the lines coincide with physical realities dictated by this tilted planet's orbit around the sun. Sharing a particular latitude, north or south of the equator, people also share a specific orientation to the sky and experience the same day lengths during each annual cycle of seasons. Identical constellation patterns span the heavens every night at the same latitude around the world. The latitude and longitude grid helps travelers know where they are, helps them navigate to destinations and recognize when they have arrived. Exploring a particular line can clarify both physical and cultural routes chosen by humanity, historically and at this moment in time.

Thirty-eight degrees north (38°N) is a temperate, middle latitude where human societies have thrived since the beginning of civilization. Between extremes of climate found farther north and south, successful agriculture and the growth of cities along the latitude depended on many ingenious solutions to the water supply problem. Altering natural water systems has consequences, though, and at this point in human history, environmental challenges associated with water are ubiquitous.

Our geographical exploration focused on stories of wetlands conservation, groundwater and river pollution, over-drafting of groundwater aquifers for agriculture and domestic supplies, dams and aqueducts to store and transport water hundreds of miles and serve distant farms and cities, and water limits generating tensions between cities, states and nations. That grim list of challenges and problems was balanced by the inspiring work of scientists, environmental activists, and resource agency employees, striving to renew what has been damaged, recover what was lost, and preserve what remains pristine.

Water is the key to life on Earth, shaping its abundance and patterns even when the influence is indirect. Lakes, rivers, and wetlands are particularly rich places of biodiversity. Adaptations to conserve water become the signature traits of desert plants and animals, and as an erosion force, "water at work" explains many landforms. Civilizations have been shaped, grown, or failed as they tapped water below ground, captured it behind dams, and moved it across the landscape to serve agriculture and humanity.

Maps depict cross-hairs where the 38°N latitude and 119°W longitude lines center on the oval expanse of Mono

Lake, east of the Sierra Nevada crest, where we worked for two decades as park rangers and still live. The lake is a salty inland sea, too alkaline for fish, but with enormous quantities of algae, brine shrimp, and flies that feed millions of birds. Picturesque tufa towers decorate the shore where springs brought calcium into the carbonate-rich lake and formed limestone spires underwater. Towers were exposed as the lake dropped after the City of Los Angeles began diverting streams from the Mono Basin in 1941. A 16-year legal battle to save the shrinking, increasingly salty lake ecosystem began in 1978, pitting concerned citizens and scientists against Los Angeles. In 1994, the David-versus-Goliath contest was won by little David. A small group of leaders had built a national organization of supporters and volunteers that prevailed in court, ensuring that the giant city shares enough water with the lake to keep it alive and healthy. There is hope in such an outcome for a world facing many similarly daunting challenges.

In 1982, the Mono Lake Tufa State Reserve was created with a single park ranger assigned to manage the new park. We, together, became that ranger, job-sharing until our retirement. It was a gratifying privilege to share the lake's story with visitors from around the world and water became the dominant subject of our ranger careers and post-retirement lives at Mono Lake. David has written several books about California water issues, while Janet still trains and coordinates the volunteers who conduct public and school tours at Mono Lake, explaining the lake's unique qualities and significant environmental history. Shaped by that background, our interest in circling the world took on a deeper meaning when we realized that the 38th parallel route intersected with so many lakes, rivers, and estuaries facing significant challenges. Our journey of discovery became a study in environmental geography.

The campaign to save Mono Lake succeeded due to a few key citizens and scientists who persevered, so we were particularly intrigued by the environmental protection campaigns encountered during our travels. In every nation and state, we found dedicated activists, educators, land managers, and local experts willing to explain their challenges

3

and accomplishments. Passionate people were the rays of hope shining on problems around the world that could otherwise seem depressingly insurmountable.

Listing major cities is only one way to describe our route along the 38°N latitude: Seoul, in Korea, Athens, Greece, and Córdoba, in southern Spain make that list, along with Louisville, Kentucky, Dodge City, Kansas, and San Francisco, California. Most of our traveling happened outside of cities, though, away from well-beaten tourist tracks. The landscapes between those points provided conditions for our most interesting discoveries. The 38°N latitude bisects the Mediterranean Sea, coincides with much of the ancient Silk Road trade route which knit together oasis towns between China and Turkey along the edges of deserts, intersects with the mouth of China's Yellow River, then finds that nation's "Mother River" again and again as the channel meanders north and south of the latitude line. It passes through the headwaters regions of the Tigris and Euphrates rivers and the Guadalquivir and Guadiana rivers of Spain and Portugal. The line finds Chesapeake Bay on the eastern edge of the United States, and the Colorado River in the west, after crossing the Great Plains along the Santa Fe Trail, the historic corridor that followed the Arkansas River upstream. The parallel also encounters islands where the surrounding sea shapes cultures and environments: the Azores, alone in the vast Atlantic; Sicily and several Greek Islands in the Mediterranean Sea, Japan's Honshu and Sado Islands, and the Farallones off the coast of California.

Within the United States, we learned about the decline of the Chesapeake Bay and San Francisco Bay ecosystems, mountaintop removal coal mining which is filling in the creeks of West Virginia, wetlands in the heart of Kansas and the over-drafted Ogallala aquifer, uranium mill tailings being removed from a bank of the Colorado River at Moab, Utah, and a quest by thirsty Las Vegas to tap into northern Nevada's groundwater, mirroring the 20th century long-distance water-grab by Los Angeles in California.

The route encountered many of the world's longest aqueducts that move water to cities from distant sources. One of the longest, shifts river water 542 miles in Turkmenistan to regional farms and to Ashgabat, the national capital. Beijing is served by a 190-mile canal tapping reservoirs on the Yellow River and plans to move additional water from farther south in the Yangtze River. Athens, Greece, imports water from reservoirs 135 miles away. Spain's Tajo-Segura aqueduct extends from the wetter north to parched southern provinces where most of the nation's growth is happening, repeating a pattern familiar to Californians. In California, the Los Angeles Aqueduct extends 338 miles from the Mono Lake basin; San Diego reaches 444 miles into northern California via the California State Water Project; and San Francisco's Hetch Hetchy System delivers Tuolumne River water 167 miles from the Sierra Nevada.

Flooding valleys with reservoirs at the upper ends of these water systems has often been controversial. In Turkey, one of the world's oldest towns, Hasankeyf, is now threatened with inundation by a dam being built on the Tigris River. In California, the early 20th century fight over whether to dam the Tuolumne River and flood Hetch Hetchy valley inside Yosemite National Park helped clarify the meaning of National Park status. Portugal's Alqueva dam has been called a "white elephant," based on a long list of concerns about its costs and actual benefits. That dams do not always, in the long-term, deliver all the forecast benefits was one of the lessons at the New Melones dam on California's Stanislaus River, where an unsuccessful battle against filling that reservoir pointed the environmental group, Friends of the River, toward many later successes.

Our route also encountered natural lakes, inland seas, and coastal lagoons that have much in common with salty Mono Lake. Saline lakes are often under-appreciated natural environments rather than "dead seas." Lagoons on the east coast of Spain support an ecosystem of shrimp, flies, and birds similar to Mono Lake, but add mud baths for people as a local attraction. *Tuz Golu* means "salt lake" in Turkish, and there companies harvest salt while migratory birds nest and feed. Turkey's largest lake is Van, with alkaline water that forms towers similar to Mono Lake's famous tufa. A national effort was ongoing in central China to save another vast alkaline sea,

Qinghai Lake, from impacts of sedimentation and stream diversions.

East of Qinghai, the city of Lanzhou was the historic junction point for northern and southern routes of the ancient Silk Road, where traders carrying goods between China and Turkey moved between oasis towns. Water, or the lack of it, dictated much of that route and shaped the history of its cities and cultures. We traced the Silk Road (and its ubiquitous camels) across western China, through Turkmenistan, and all the way through Turkey.

In Korea, after venturing inside the DMZ, we learned about the renewal of an urban river through downtown Seoul and the much more controversial "Four Rivers Restoration Project" that is deepening and channelizing the banks of the Han River, upstream from Seoul, carving away natural wetlands and displacing rice farmers.

Wherever wetlands exist, wildlife populations are particularly rich. Where environmental manipulation damages natural systems, protecting and restoring habitat for endangered species becomes critically important. We visited captive breeding programs for Japanese Crested Ibis and for Iberian Lynx in Spain and learned about endangered seals and sea turtles on Greek islands. Wetlands, once thought lost forever, were being restored in San Francisco Bay, the Sacramento-San Joaquin Delta, and in Point Reyes National Seashore to benefit beleaguered species of fish, birds, and amphibians.

At several places where water and land funnel migrating birds into narrow corridors, we joined with volunteers and scientists in annual counts of raptors and songbirds. Because locations on the 38th parallel are midway between the arctic and the tropics, they experience four distinct seasons. North-south migration routes cross latitudes to seek "endless summer," or at least avoid the harshest effects of winter. We joined bird-counters at Point Reyes, near the Golden Gate on California's coast, at Chesapeake Bay on the east coast of the United States, and in Sicily, where the count began as a campaign to end an annual slaughter of migrating raptors.

The woman behind the Sicilian campaign endured threats of violence as she built an international monitoring effort. Her inspiring success story was one of many we encountered – stories of dedication and struggle that enriched our journeys. The list of people who took time to meet with and educate us grew amazingly long. Acknowledgments in the back of this book include those names along with websites from the associated agencies, environmental groups, and campaigns that are striving to protect communities, lands, and local cultures.

North or south of the 38th parallel, few other global latitudes pass through so many states and nations. Because of land mass and population concentrations along the 38th parallel, the Robinson map projection system, now widely used for world maps (adopted by the National Geographic Society in 1988), selects 38°North and 38°South as "standard parallels" – the two lines across the map that depict size and shape most accurately, while distortions occur at all other latitudes.

Following a straight line in almost any direction all the way around the world makes comparisons and connections between places more evident. Other latitudes must also provide meaningful insights, but we encountered an intriguingly long list of stories within a half-degree either side of the 38°N line. We only strayed farther afield when travel logistics made it necessary or, in a few cases, to reach parts of an extended watershed that clarified the story of a river or lake along our route.

Though this narrative proceeds westward around the globe as through describing a single journey, we actually spread the traveling across three years in separate trips to Europe, Asia, and across the United States. We traveled by bus, train, or rental car, and flew between countries and across some of the long distance segments within China, and we trekked on foot, bicycles, and by boat across California, on a trip that put all of the global experiences into perspective.

This adventure grew out of an interest in geography that seems hard-wired into most humans and – at the age we learn that the Earth is a round ball – often manifests itself in questions like, "Where would I end up if I dug through to the other side of the Earth?" Many of us know the pull of the unknown, the road not (yet) taken, the attractive mysteries in the unlabeled "white" places on the map.

In recent years, global positioning system (GPS) devices have become popular navigation tools for hikers and drivers. The Google Earth website provides satellite imagery of the entire world referenced to latitude and longitude. Such widespread interest in navigation suggests that other people may want to visit the locations we describe, or may be inspired to explore other latitudes in similar detail. For this book, we record latitude and longitude coordinates in the classic degrees-minutes mode (38°30'N), rather than digital equivalents (38.5°N). Throughout our travels we posted updates at http://paralleluniverse38n.blogspot.com, with hundreds of additional color photographs and links to websites that may help people who want to follow in our footsteps.

Each travel segment was carefully planned, yet there were unanticipated encounters, a few pulse-raising adventures, and events that forced us to alter our travel plans. An ash cloud from a volcanic eruption in Iceland enveloped northern Europe and stopped air travel while we were on the Azores Islands. Nationwide strikes in Greece closed the airport the day we were to fly to Athens. The only buses going to our destinations in Turkey were sometimes fully booked due to school holidays. Planes were often late. Taxi drivers became lost. Kind strangers often helped us find our way. Successful travel is about serendipity, flexibility, and acceptance.

Four weeks before our planned trip to Asia, the Earth shook with incredible force on the morning of March 11, 2011. The magnitude 9.0 earthquake had its epicenter at 38°19'N; 142°23'E, about 60 miles offshore from Japan's largest island, Honshu. The tsunami that struck the east coast minutes after the initial quake caused the most tragic destruction. Entire coastal villages were demolished. Over 21,000 people died and a half-million were displaced. The quake damaged several nuclear power plants where reactors lost cooling capability and melted down, releasing radioactivity to the atmosphere and the sea. The nation struggled with body recovery and critical shortages of food, water, and medical supplies. It was not an appropriate time to visit the 38th parallel in Japan.

After six months of reconstruction, we finally did visit and were able to write the final chapters of this book. Efforts toward revival and recovery since the tragedy were exceptional and inspiring. A desire to renew conditions for abundant life, wherever they have been diminished or destroyed, was a trait we saw manifested everywhere, around the world, on the 38th parallel.

Wendell Berry wrote that "nobody can discover the world for anybody else." We traveled the 38th parallel to put our feet in the Tigris and Euphrates rivers that served so much of early human history; to personally experience the dense press of humanity where China's Yellow River has nourished and challenged its residents for thousands of years; to note the remnants of historic waterwheels and aqueducts and cisterns across southern Europe; to compare the chill air of central Asian glacial valleys to the summer monsoon of South Korea and feel the changes in weather and humidity as we left the verdant eastern states of our home nation and returned to the arid west. Our discoveries are shared here because the written impressions and the voices of people we met along the way can be a first step toward your personal experience, toward a common bond that values a long-term, sustainable relationship with water around the world.

UPDATE 2020
The hardcover edition of this book was published by UC Press with 30 black-and-white photos. We are pleased to provide a paperback edition with 64 color photos to better reflect our vividly colorful memories. Most of the issues we documented still need resolution and all have been exacerbated by the global climate crisis. Human emissions are warming the atmosphere and oceans, delivering impacts though the planet's water cycle. This has become our most urgent environmental challenge.

Part I: Asia

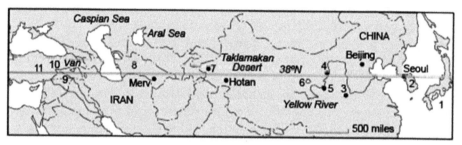

1. Japan 2. South Korea 3. Xi'an 4. Yinchuan 5. Lanzhou
6. Qinghai Lake 7. Kashgar 8. Turkmenistan 9. Tigris River
10. Euphrates River 11. Cappadocia, Turkey

"What we have to work on in the 21st century is to overcome the division between people and the environment, so the future of humans won't be different from the future of nature."
A South Korean high school student

Dams and Dredging: the Four Rivers Restoration Project

Our journey began in Korea. Hundreds of years ago, Seoul, the capital of modern South Korea, was a newly founded village along the banks of a pretty creek called Chonggyecheon (37°35'N). As the city grew, the creek became a sewer and finally was covered over by concrete and a freeway. Several years ago, Mayor Lee Myung-bak decided to bring the creek back to the daylight. Now, a semi-natural stream park serves the urban residents of the city. We walked there from our lodge, a traditional inn with bamboo walls, sliding doorways, and an ethernet connected computer in every room; modern Seoul is one of the world's most "wired" nations.

Walkways line both banks of the flowing creek, which is punctuated by cascades and stepping-stone bridges. Chonggyecheon impressed us as an urban park, though water from the Han River must be pumped at considerable energy costs to enhance the flow. The park is used day and night and clearly appreciated by local residents, from splashing children, to romantic couples strolling hand in hand, to elderly dog walkers. The mayor moved on to become the nation's president and his new campaign was to redesign South Korea's four largest rivers – the Han, Nakdong, Geum, and Yeongsan – a much more controversial objective.

In 2009, gigantic excavators began carving riverside bluffs away to double the width of the channels while digging them 12 to 18 feet deeper. The $20 billion national project was constructing 16 new dams on the main channels of the four rivers, plus five more on their tributaries, while enlarging 87 existing small dams and armoring over 200 miles of riverbanks. President Lee called the effort part of a "Green New Deal," intended to store water against the prospects of drought, to prevent flooding, to improve water quality, restore river ecosystems, promote river-related recreation, and (perhaps, above all) to stimulate the economy through 190,000 construction jobs and spending that equaled almost 20 percent of the nation's gross domestic product.

The Chonggyecheon River has been resurrected through downtown Seoul

The goals of the Four Rivers Restoration Project sounded commendable, but when viewed closely, the list of benefits seemed exaggerated. We learned the details and toured part of the construction with the Korean Federation for Environmental Movement (KFEM). Their national headquarters is a comfortable old residence converted to offices, with commuter bicycles lining the walls. There we met with Choony Kim, the organization's Chief of International Affairs.

South Korea, though densely populated, has plenty of water; episodes of flooding occur primarily on upper tributaries, rather than the main river channels where the work was focused; and the engineering approach will likely degrade, rather than improve water quality, because slower water movement increases accumulations of algae and pollutants. Most of all, the removal of natural wetlands and streamside vegetation is clearly not "river ecosystem restoration," but rather destruction of habitat. Natural beauty and wetlands critical to migratory birds and other wildlife were being replaced with bike paths, spraying fountains, and many miles of city-style parks.

When we pointed out that dams and channelization are old ideas about how to manage rivers, Choony said, "Many experts agree." She had worked for KFEM since 1995, after studying at the Yale School of Forestry in the United States. She speaks excellent English and was eager to talk to international visitors about the project, to balance publicity that the government was providing to media outside Korea. Choony showed us dramatic before-and-after photographs taken at construction sites. Because the rivers are being widened, villages and farms must be relocated. Her frustration showed as she explained that some of the displaced rice farmers, following years of encouragement by the government and environmental groups, had transitioned to organic farming, controlling insect pests in the paddies with ducks instead of pesticides and using weed-eating snails in place of herbicides. Shutting down those farms near the river meant 15 years of effort wasted.

An unexpected impact from so much lost farmland was a national "kimchi crisis" in the autumn after our visit, as a shortage of cabbage was blamed on lost farm production due to the Four Rivers Project. Kimchi, the spicy national dish, was served at every meal we ate in Korea. The cabbage shortage became a political issue, debated by the national legislature.

"The government calls this 'green economy,' but has no concern about the ecology," Choony said. "They just keep construction workers busy, busy, busy." The project is being pushed along very fast because South Korea's presidents are limited to one five-year term, so President Lee "wants everything done in his time," Choony said.

The river-widening and excavations do face opposition within the country. Polls find that the majority of the public opposes the work, both for its $20 billion cost and ecological impacts. In a national election just before we visited, the President's ruling party had lost about half of their control in the legislature to opponents from other parties. Despite that change, the project was forging ahead, with a goal to finish by 2013. We asked Choony what KFEM hoped would happen to successfully stop the work.

"The president cares about voices from the international community. He took part in the G20 meeting in Toronto, Canada. Korea will be the host government next time. He has created the Global Green Growth Institute. They want to advocate this Four Rivers Restoration Project to the international community, saying it creates jobs and 'green economic growth.' We say, "No, if you look at this case, it's not real.' We give out information, but other nations listen to the government. So, we need to give information to the international community, because he may listen to that voice. The international voice can come back to the domestic voice."

In the United States, a project of this type would require major environmental impact studies and possibly be limited by lawsuits. In South Korea, KFEM lost a lawsuit against the project based on its impact to an endangered aster plant species that depends on the riverbank environments, because the government promised to mitigate impacts by rescuing the asters from excavation areas and cultivating them in special plots.

From a watershed that gathers runoff from mountains to the east and southeast, the great Han River passes through Seoul, then turns northwest until it enters the ocean at the Demilitarized Zone. We followed the Han River upstream from Seoul to Yeoju, a city on the south fork, traveling in a KFEM photographer's car. Jong-Hak Park had a medallion hung on his dashboard showing that he supported the International Crane Foundation. Now 68 years old, he had worked with KFEM for 12 years, and told us, "I love KFEM, because my grandson has allergy. I heard that it's from environmental pollution, so I will keep the environment."

Two more young and energetic KFEM staffers, Yong-un Ma, the organization's wetlands expert, and Naree Jeong, coordinator of a Nature Conservation Team, showed us a six-mile stretch of the river where construction crews were building three dams and excavators filled trucks with loads of riverbed sand and gravel, adding to mountains of material piled nearby. The scale of the work was shocking. We had never seen so many gigantic excavators at work, digging,

14

widening, chewing up and spitting out the former riverbed. The dams we saw under construction were close enough together so that a series of excavated "water basins," separated by gates and locks, could conceivably become a continuous canal to serve shipping. Opponents of the project think that is exactly the point, as a grand canal connecting the nation's rivers had been a key objective for President Lee when he campaigned for office, until negative public opinion stopped that plan.

"Four Rivers Restoration Project" widening and deepening the Han River

The Koreans working against the project were inspirational. It seemed hopeless, yet they were bearing witness, talking to anyone who would listen, asking us to "please tell the rest of the world what is happening to our rivers." KFEM was created when several South Korean groups involved in anti-pollution and anti-nuclear issues merged in

1987. The non-governmental organization has fifty local chapters, whose 30,000 members pay monthly dues.

We were treated grandly that evening, with a dinner feast at a restaurant overlooking the river, owned by Hang Jin Lee, the head of the Yeoju chapter of KFEM. He told us that he was "not a born environmentalist, but about 10 years ago my friends asked me to work for the environment. I began to realize that environmental protection is very, very important. And now I cannot leave this work." He had given considerable thought to two opposing characteristics of humanity, consciousness and greed, hoping "we can regain the essence of consciousness – the relationship with living things and non-living things."

From the window in our hotel room, that night, we had a view of the Han River, midway between the two work sites we had seen that day, where excavators continued to chew away at the riverbank.

The United States went through a river-damming and channel-straightening/concrete-armoring stage in the last century. We learned that a more effective way to manage watershed systems is to allow floodplains to absorb high water episodes, which regulates flows while the riverbank vegetation filters impurities. So, it was frustrating to watch the living river bottom being pulled out and lush riverside plateaus being excavated. Though the momentum of the work seemed impossible to stop, opponents of the project remained determined. On a poster in the Yeoju KFEM office, an excavator was depicted being restrained by the endangered aster plant. Mr. Ma told us the words on the poster read: "Be Persistent; It is Your River."

UPDATE 2020

During the 2012 Ramsar Convention on Wetlands conference, the Four Rivers Project was given the "Grey Globe Award," a dubious distinction for policies and projects that destroy wetlands. After President Lee's term ended in 2013, the new administration began investigations into the Four Rivers Project. A "Citizens' Committee for Returning the Four Rivers to Their Natural State" formed. In April 2018, a weir on one of the Han River tributaries was dismantled to restore water quality and allow free flow, serving as a model for mitigating the damage caused by the former administration's disastrous river project. In June 2019, the Citizens' Committee announced that 20 NGOs from Europe, Africa, North America, and Asia had jointly released a statement of support for the current government's intent to dismantle weirs on the four rivers to restore them to their natural states.

The Korean DMZ, Ecological Recovery Behind Barbed Wire

On the 60th anniversary of the start of the Korean War, June 25, 2010, we entered the Demilitarized Zone (DMZ) that separates North and South Korea. We were with a delegation from Pasadena, which has a "Friendship City" relationship with Paju, a booming South Korean city on the edge of the DMZ (37°46'N). Few travelers are allowed inside the Joint Security Area of the DMZ, administered by the United Nations, so this was a unique opportunity for our 38th parallel exploration.

The DMZ is a 150-mile long and 2.5-mile wide strip of land across the Korean Peninsula coinciding, more or less, with the 38th parallel. A peace treaty was never formalized to end the Korean War, and North and South Korea each keep hundreds of thousands of soldiers massed along their borders in a tense military stand-off. The war began in 1950, the year David was born, and the cease-fire came in 1953, Janet's birth year. During our entire lives, soldiers have stood watch over a land fringed with barbed wire and perhaps a million landmines, while the land healed itself and an inadvertent ecological recovery occurred.

Most of that landscape was farmed for thousands of years, then ravaged when it became a battlefield. Ironically, after years of war, 1,100 types of plants now grow behind the barbed wire and the land is green and lush. At the western edge of the DMZ, the Han and Imjin Rivers estuary provides rich tidelands and riparian habitat for wildlife, including 18 endemic fish species (only found there) and the endangered Korean Golden Frog (*Pelophylax chosenicus*). Many endangered or threatened bird and mammal species are thriving, including three species of cranes that migrate in each winter. Asiatic Black Bears (*Ursus thibetanus*) are inside the DMZ, as well as leopards, lynx, and, perhaps, the exceedingly rare Siberian Tiger (*Panthera tigris altaica*), if glimpses seen on military surveillance cameras and territorial scratch marks on trees are being correctly interpreted.

Guards in the Demilitarized Zone look at North Korea across the Imjin River

Winter is the peak season for migratory birds; but summer residents we saw included herons, egrets, and magpies, flocks of Bean Geese (*Anser fabalis*), and the colorful Black-capped Kingfisher (*Halcyon pileata*). We had numerous fleeting looks at Water Deer (*Hydropotes inermis*), whose males have large tusks at their jaws, but no antlers.

Seeing any wildlife was a treat, as we were only allowed to travel directly to Daesong-dong village, where 212 South Koreans live inside the DMZ (37°57'N). The Pasadena delegation was granted special permission to enter the Zone to dedicate a new dental clinic there on that 60th anniversary of the war. The group included Dr. Yung Nam, a Korean-American dentist, who was the inspiring force behind the clinic. Until this time, we had no idea that *any* people actually lived inside the DMZ, but both South and North Korea are allowed small "unification" or "peace" villages. The rest of the Zone remains unoccupied, and that is where the ecological experiment in human exclusion has been underway.

Our group passed through checkpoints where serious-looking soldiers closely examined our passports and paperwork before giving the driver a piece of cloth to hang from the window of our van – a low-tech signal that we had permission to be inside the Joint Security Area itself. Our initial impressions, beyond the checkpoint, were of verdant forest and quiet. The energy and bustle of modern South Korea had been left behind.

Dr. Nam's plans to dedicate the dental clinic at Daesong-dong were thwarted, at the last minute, because of tension with North Korea over the torpedoing of a South Korean gunboat. United Nations administrators were concerned that a ceremonial gathering, even for such a benevolent purpose, might be considered provocative by North Korea. It was a reminder of just how tense (and sometimes illogical) the relationship remains in that region.

Though the clinic dedication was off, our delegation was still allowed to visit the school where it will be someday be established. From a rooftop observatory in Daesong-dong, our eyes were inevitably drawn toward structures in the North Korean "peace" village just 400 yards away. A North Korean flag flew from a provocatively tall pole, an invitation to a flagpole "arms race" that the South has declined.

Our guides pointed toward hillsides denuded by the fuel-wood scavenging of impoverished North Koreans, where deforestation has caused severe floods The southern side's thick forests were a striking contrast, though Dr. Nam recalled barren hills around his home near Paju when he was growing up, during and just following the Korean War. After the war, the South Korean government put many citizens to work planting trees to reestablish its forests.

The delegation was lodged in Paju City, where business cards we were given by officials included a slogan (also mounted on every taxi in the city): "G&G Paju," standing for "Good and Great Paju." Undeterred by the tensions of the DMZ that it borders, Paju is one of the fastest growing cities in fast-growing South Korea. Paju's population increased by over 80 percent in the prior decade, and the mayor's office had rather incredible plans to double the population in the coming decade by attracting high-tech industries. This frenzied economic activity is a concern for people watching the ecological recovery within the neighboring DMZ and wondering what the future will hold, if and when that land is once again opened to development.

The DMZ Forum, with headquarters in New York, was founded in 1994 to promote the transformation of the DMZ, a symbol of war, into "a place of peace among humans and between humans and nature." They hope to build enough international support for a nature preserve to forestall the destructive development impacts if the Zone ever opens to the world.

The local DMZ Ecological Research Institute works with South Korean school children to shape such attitudes within the nation itself, by taking students into the Civilian Control Zone (CCZ), a strip of land three miles wide that parallels the southern edge of the DMZ. Though also fringed with barbed-wire and closely guarded, the CCZ is not barred to South Koreans. Some commute there to work on family farms, while many join international tourists touring Korean War historic sites. Our guide was JungRok An, a student at the Korea University in Seoul, who founded the Youth Exploration Team program for the Institute that brings high school students from all around Korea.

Though only 20 years old, JungRok is an eloquent and knowledgeable researcher, dedicated proponent of environmental education, and born leader, who told us: "To protect the DMZ for the future, I think youth are the most effective." "EcoYouth" students receive one year of classroom orientation and attend field camps to learn research techniques, then begin individual or small group ecological study projects within the CCZ. JungRok's own involvement began five years earlier with research on Swan Geese (*Anser cygnoides*). His father heads the Education section for the Institute.

21

We were with a group of Youth Exploration Team students from high schools in Seoul on their first day of orientation. The DMZ Ecological Research Institute's education staff handled two other groups that same day: high school students who were a full year into their studies, plus an elementary school class.

Though South Koreans can visit the Civilian Control Zone, these students had never been there. Our itinerary included the Dorasan train station where South Korean trains must stop, though tracks continue to the north. Map displays explain how ties to Russia and Europe will become possible whenever the route opens. We also went to places tourists never see, including wetlands along the Imjin River.

Along the Han and Imjin Rivers, local farmers from Paju grow rice and ginseng. Loosely strung electric lines serve irrigation pumps that lift river water to the fields. Ginseng has become a key money-maker for farmers, but requires plastic covers on frames to shade the plants. The covers keep out birds and other wildlife, so the continuing expansion of ginseng fields concerns JungRok. Rice fields, by contrast, are beneficial feeding and resting wetlands for migrating waterfowl (a role we were familiar with from the Sacramento Valley and Delta of California, where the stubble left in flooded rice fields after harvests serves seasonal bird migrations). Along   the roadsides, dozens of red triangular signs with skulls-and-crossbones warned of land mines. The kids found the grim warnings fascinating and took photos of each other posed in front of the signs. At one point, as we got out of our van to look at a pond, soldiers on patrol told us we had to move along; that we were not supposed to stop there. Reminders of armed tension were unavoidable.

That afternoon a drizzling rain began to fall. The annual summer monsoon was forecast to start that day, and it did, at 4 PM, right on schedule. The highway back toward Paju paralleled the river estuary, where egrets fished behind chain-link fence topped by razor-wire.

We asked JungRok what he most wanted people to realize about his project and the Institute: "Before I started these DMZ ecological exploration teams, I thought "What can I do to help this environment that is now endangered by humans? One way to really save this environment is to change the mind itself, so that's why I thought of youth. The whole environmental problem is on us, the youth, because it will happen when we are older, and it will happen for our children. So, if youth are talking to adults about this environment and how it is beautiful and mystical and important, it could bring change."

On the van ride out of the CCZ, we rode with six high school students who had already completed a year of work. They came armed with nets and collecting jars and had spent the day sampling water quality in ponds and documenting populations of aquatic invertebrates and amphibians. They were eager to speak into our tape recorder, as if it were a direct audio link to the world, to explain what they wished people to know about the DMZ and their hopes for the future.

"Hello listeners," one girl began. Janet told her that she would someday be a television personality or a diplomat, which made all of the students laugh. "I hope you know that the DMZ in Korea is a very valuable, ecologically important place. Even though the 'demilitarized' name is a meaning about military values, there are also environmentally important things in there, and even after Korea is unified, we need to keep up the habitat interests of the place. Thank you."

A quiet young man from CheongShim International Academy provided this eloquent conclusion: "I think you two just realized, today, what I realized as I was working as a DMZ member. You mentioned the irony you found here, how the war between people led to a better future for this environment. I think that's the main point. What we have to work on in the 21st century is to overcome that irony, the division between people and the environment, so the future of humans won't be different from the future of nature."

UPDATE 2020

Peace has not been achieved between North and South Korea, so the DMZ still remains an untouched wildlife sanctuary. JungRok An entered the Harvard University Graduate School of Design in 2016 and has now returned home. In 2019, UNESCO designated parts of the CCZ as "biosphere reserves," hoping that eco-tourism might lessen local resistance to environmental projects and pressure to build roads there.

China's Yellow River Delta

High-speed freeways paralleled the east coast on the long drive from Beijing, China, south toward the city of Dong Ying. Orchards were blooming (it was mid-April) and rows of cottonwood trees along the highway were leafing out, recently planted as part of a massive national campaign to capture atmospheric carbon; but also, as we would see in central and western China, as a "green wall" against expanding deserts. Many millions of trees had already been planted.

Our guide and translator across eastern China was Kinder (pronounced "kin-dur") Jinde Shu, who has a special interest in natural places, migratory birds, and helpful connections with China's parks and nature reserves. He was China's first member of the National Association for Interpretation (NAI), an American organization for professionals who conduct programs and design exhibits for parks. He has translated NAI materials into Chinese and coordinated training for China's parks and reserve staff. Our starting point on the 38th parallel in China would be the National Nature Reserve at the delta of the Yellow River, where a special tour had been arranged by the Deputy-director, a personal friend of Kinder's.

The highway repeatedly crossed over water channels along a very flat landscape. Such terrain, along with the extremely high silt loads the Yellow River carries, explains why the river (the *Huang Ho*, in Mandarin Chinese) repeatedly fills its river channel with silt and then jumps into new routes to the Bohai Sea, a northern gulf on the Yellow Sea. The mouth of the Yellow River has shifted more than 50 times in the last 150 years. Today the Yellow River enters the ocean several miles south of the 38th parallel, but just a few years ago it occupied a channel exactly on the latitude line.

This river carries the highest silt content in the world, a brownish-yellow sediment-load that explains its name ("*Huang*" means "yellow") and averages 1.049 billion tons of material each year. "When the Yellow River flows clear" is a Chinese saying for an event that will never happen (similar to our expression "when pigs fly.") The delta zone adds an average of 8,000 acres annually that keeps expanding China's youngest wetland.

The Yellow River flows west-to-east, beginning as snowmelt in the Kunlun mountains of Tibet, and passing through nine Chinese provinces. The 3,400-mile-long river's mainstream and its 35 tributaries have a combined length totaling 8,380 miles across a 290,500 square mile watershed. It is China's second longest river (the Yangtze, about 500 miles longer, follows a parallel path farther south). Our four-week trip across China began at the delta and would encounter the Yellow River several times more, far inland.

From our perspective traveling upriver from the coast, the river channel bent southwest though alluvial plains of wind-blown silt ("loess"), then turned north for hundreds of miles (circling the Ordos Desert within Inner Mongolia), before making a great sweep south again, then southwest toward Tibetan headwaters. In central China we would follow the main course of the river for about 250 miles and then along a tributary for 100 miles, the Huang Shui, which descends from eastern Qinghai Province.

Where the Yellow River crosses flat floodplains it has been confined between levees for thousands of years. Sediments inexorably settle onto the riverbed, forcing residents to keep elevating flood-protection walls in response. In places, the river runs 20 to 30 feet *above* adjacent farmlands and villages, putting the population at tremendous risk if levees fail. The Yellow is not only China's "Mother River," sustaining life along its long run from mountain snowfields in Tibet, but it is also the nation's "River of Sorrow." Archaeological evidence of disastrous floods dates back more than 2000 years. In 1887, one of many historic Yellow River floods killed nearly two million people. The worst flood disaster in world history occurred here in August, 1931, killing about 3.7 million people. Such numbers are worth a moment's reflection;

China's vast landscape and population size make the scale of disaster events unimaginable.

The Mother River has, ironically, also been given very little respect, as it became a convenient sewer and waste disposal point for towns and factories. Some segments of the Yellow River are so badly polluted that the water is even unfit for farm irrigation.

From the bustling industrial city of Dong Ying, we drove about 40 miles to the delta. It was hard to believe a protected wetland would be at the end of that road. Construction zones gave way to farmland, then, China's second largest oilfield, and, finally, to marshes. At the gate of the Yellow River Delta National Nature Reserve, a massive Oriental White Stork (*Ciconia ciconia boyciana*), lifted off in a flash of black and white. Twenty-eight pairs of storks had bred at the delta that year, which serves as an avian "international airport" each autumn, winter, and spring. Common Terns (*Sterna hirundo*) streaked by in a high wind as we walked down boardwalks with intricately carved handrails to elevated birdwatching towers.

Our Reserve guide was Shen Kai, a young engineer on the staff and a bird expert who carried a camera with a huge telephoto lens. His supercharged energy kept us zooming between spots in the wetlands, struggling to keep his car in sight. The Reserve was created in 1992 to enhance and protect the delta and its wetlands for migratory and nesting birds. Over 6 million birds use the protected lands every year: 296 species, including six of the world's nine species of cranes. Seven of the species that depend on the Reserve's wetlands are "class 1 priority" species in China (equivalent to the United States' endangered species designation), including cranes, eagles, and storks.

Two separate land sections, together totaling 378,000 acres, reflect the most recent major shift in the river's path: a northern Reserve parcel coincides with the Yellow River's sea entrance before 1976 (on the 38th parallel), while the present channel passes through a larger land parcel about 15 miles to the south. Huge oil and natural gas fields lie between the two parcels, with some oil wells pumping inside the Reserve itself. Pollution is a concern, along with habitat fragmentation caused by roads and installations, but the managers' greatest struggle is to secure enough "eco-water" from the main river to optimize conditions in the marshes.

With increasing water consumption upstream, the Yellow River only reached the sea one year in the 1990s. Like our country's Colorado River, which often dries up before reaching the Gulf of California, there have been too many competing demands for human use of the Yellow River's water. The national Yellow River Conservancy Commission (YRCC) responded to that problem by regulating water use, setting specific allotments for each of the nine provinces along the river basin. Amounts were based on an estimate, made in 1987, that the river carried 15 trillion gallons of water annually, but – in another parallel to our Colorado River, where flow rates were over-estimated when states negotiated water rights – in recent years the Yellow River's volume fell to 14 trillion gallons; in 2003 flows were below 12 trillion gallons. We visited during the spring dry season, yet flowing water was reaching the sea, thanks to the new conservation regulations of the YRCC. Since 2003, there has been enough reliable flow to allow water diversions into wetland ponds, "but still only half of what that landscape needs," according to Shen.

The rainy season here arrives with the hot days of the summer monsoon. Flow in the Yellow River peaks between July and October, fed by heavy monsoonal rains across the eastern half of China. The summer months, between May and October, also bring more than 150,000 tourists. Spring is dry and windy, as we can attest, as morning winds nearly swept us off our feet while we stood on the rooftop viewing platform of a new visitor center.

The modernistic building, completed in 2010, looms over the landscape, six stories high. It is called "the Ecological Port," because it sits on the south bank of the river where boat tours begin. From the rooftop we had jaw-dropping views of the marsh and the Yellow River heading toward the ocean, just a quarter-mile away. There were not yet any services inside the new center – all funding seemed to have gone into the incredible structure – but it appeared large enough to handle thousands of visitors. On less windy days, two tour boats were available for cruising the river out into the nearby sea. A dramatic mixing-line between blue ocean and "yellow" river water is a "must-see" sight that the wind kept us from viewing.

China's Yellow River is colored by the heavy loads of silt it carries

The howling gusts, thankfully, stopped during our lunch in nearby Xian He (the town's name means "holy crane"). We were honored with a special meal arranged by Kinder's friend, Li Juanzhang, the deputy director of the Reserve. Mr. Li was unable to join us, but we feasted on unfamiliar fish, sea cucumber and octopus dishes; far too much food for us and our guides to finish. As we had experienced in Korea, the culture calls for excess when honoring guests.

During lunch, Shen Kai told us that what he liked best about working at the Reserve was protecting the birds and the special character of the wetlands and tidal lands. He grew up in the local area and had a university degree in petroleum engineering.

That afternoon, with the air blessedly calm, we watched Red-crowned Cranes (*Grus japonensis*) in flight. We had missed the breeding season for cranes, so most of them had migrated north again, but some were kept in cages and turned loose daily to fly free. They circled near us, again and again, in one of our most memorable experiences of that day. Cranes are the "birds of heaven," the "holy bird" of China, associated with love, happiness, and all good things, especially longevity. Their lifespans approach eight decades (the International Crane Foundation in Baraboo, Wisconsin, documented an 83-year-old Siberian Crane). Often, pine trees and cranes are depicted together in Chinese art as both are long-lived.

When the practice of releasing captive birds to fly free began, some birds, inevitably, chose to not return, despite the lure of abundant food. Mr. Li told us that was fine. "It was best for them. But when they do come back, it shows good relations between cranes and people."

That philosophical opinion was expressed during our evening dinner, another feast hosted by the Reserve. Kinder had met the Deputy-director when they worked together on a Siberian Crane Wetlands Project. Twenty years earlier, when the first protective status began at the delta as a local effort, Mr. Li began working there. Within a few years, in 1992, the Reserve became a National Nature Reserve.

Beside each dinner plate that evening was a small toasting glass that a waitress skillfully kept full to the very brim with a rice liquor. There were many toasts, including one to adequate "eco-water" for the Reserve. As Kinder translated, Mr. Li told us how difficult water management can be, but he added that serving the birds was a worthwhile, satisfying goal.

"From the view of conservation management, it is a challenge," he said. "Personally, it is a very heavy burden to maintain this piece of land, but it is an important job. Compared with the United States and other developed nations, where it is easier to preserve nature with established policies and strong public involvement, we have much to improve. Fortunately, we have a supportive local government, so that gives us a lighter burden." For example, after applying to the central government for $5 million for wetlands restoration, state and local governments provided another $12 million. The regional wealth based on oil and natural gas industries made that generosity possible.

The Colorado River and Yellow River have much in

Red-crowned Cranes migrate annually to the Yellow River delta

common. Both drain vast regions within each country and were named for the colors given by heavy loads of silt ("*colorado*" means "red" in Spanish). In both cases, so much water was diverted that the rivers stopped reaching the sea. Perhaps, we agreed, there should be a "sister rivers" relationship established between the two.

Though ecological health requires more water than the Reserve has yet been able to secure, it was good to see that the infamous notoriety of a dry Yellow River delta had, at least, been addressed. Regulations imposed by the YRCC the entire length of the watershed show that China can respond to the tensions between rapid industrialization, historic water uses by cities and farms, and the natural environment. Whether national regulations are actually being implemented in the distant provinces was a question we would explore further as we crossed the country. We left the Yellow River delta reassured by the enthusiasm of the Reserve staff and hopeful that China's Mother River might continue to reach the sea with "birds of heaven" winging overhead.

The South-North Water Transfer Project

As we drove back toward Beijing, our route crossed the path of a new aqueduct being built to bring water from the Yangtze River to Beijing, but it was underground, out of sight. The first phases of a gigantic South-North Water Transfer project were being constructed to shift water from southern China, where 80 percent of the nation's precipitation falls, to the drier north, particularly to address the thirsts of industry and population growth near Beijing. While the imbalanced relationship between wet regions and human thirsts is reversed in California, where *north*-to-*south* aqueducts serve Southern California and agriculture, the approach to water supply issues is the same: build long-distance aqueducts. China intends to complete 1,800 miles of canals along three routes to ultimately move almost 12 trillion gallons of water each year. Work began in 2002 on the middle route from the Han River and the eastern route to bring Yangtze River water. They hoped to finish the eastern aqueduct before China hosted the 2008 Summer Olympics, but missed that goal (completion is now projected to be in 2013). The third, western, route would begin high on the Tibetan plateau at the headwaters of the Yangtze, and faces even more daunting engineering challenges, so work there has been delayed.

The route of the eastern aqueduct uses existing parts of China's ancient Grand Canal. That feat of early engineering began 2,000 years ago, continued by successive dynasties, to make barge travel possible along 1,100 miles of canal connecting the Yangtze, Yellow, and other rivers (finally linking Hangzhou, in the south, with Beijing in 600 A.D.). Perhaps 2.5 million laborers worked on the canal and a paved highway along its length. The 21st century construction is all about water supply, rather than transportation.

Water coming from the Yangtze is very polluted, so significant costs for the project include construction of hundreds of new sewage treatment plants in the north. China's national water-needs increase step-by-step with the country's recent phenomenal economic growth, as new industries, especially expanding energy production, require more of the nation's already limited water supply. The Chinese have pioneered coal power plant techniques that use much less water, yet, overall, thirst still grows while the available water supply is actually declining, because climate warming is altering rainfall patterns and melting glaciers. A Circle of Blue series of reports, titled "Choke Point China," analyzed conflicts between water and energy and noted: "China's total water resource, according to the National Bureau of Statistics, has dropped 13 percent since the start of the century. as much water lost to China each year as flows through the mouth of the Mississippi River in nine months." (Schneider, 2011).

We were returning to Beijing to catch a flight the next evening. Janet had been to that city before and thought the air was cleaner than the last time she visited in the summer of 2007. Air pollution was so bad then that she could not see buildings a few blocks away. The difference might simply have been seasonal; we were traveling, this time, in the spring instead of summer. But major efforts had been made to clean up the air for the 2008 Olympic Games and Kinder told us that several polluting factories and power plants, shut down that year, had never reopened. Still, smog, particularly small particulates from coal-fired power plants, truck and car exhausts, and dust, remains a major problem in China's cities. After walking on the streets in each city we stayed, our throats and eyes would be irritated.

China is trying to address the problem with cutting-edge technologies to reduce coal plant emissions and by encouraging cleaner vehicles. Several of the cars and vans we traveled in were fueled with compressed natural gas. Small three-wheeled electric trucks and electric motorcycles were common; often the motorcycles were parked on sidewalks with extension cords running inside stores to charge batteries. The national tree planting campaign was another way to clean air and reduce dust pollution.

Kinder had worked on an International Friendship Forest in conjunction with the Olympics, planted in the shadow of the Great Wall, north of Beijing. During our layover in the city, he was eager to show us the native plantings, extensive trail system, and interpretative signs. Unfortunately, a guard was keeping people off the trail that day, due to fire danger, and the start of the path had been co-opted as a motor scooter parking lot. Dust and sun-fading made his signs hard to read. Kinder was obviously disappointed with the upkeep.

This pattern was repeated at other national parks we saw in China: money and thought poured into a facility without long-term operational support. In most parks we saw grandiose buildings, fancy entrance gates and fountains, but too few maps, brochures, or information panels for visitors.

Kinder may be the person to correct that situation. Feeling it is important that Chinese national parks and nature reserves provide meaningful science and conservation stories behind the scenery, he hopes to organize more training for park staff, including international exchanges with park interpreters.

From Beijing, we flew southwest to the transportation hub of Xi'an, where we took a rainy-day tour of the city's old walls and joined tourist hordes to view the famous Terracotta Warriors. Xi'an and the Great Wall near Beijing were two places in China where we did not stand out as the only Westerners.

Xi'an had been the eastern starting point for the historic Silk Road, a trade route we would follow for hundreds of miles in western China. It was the capital of the Han Dynasty during the peak of Silk Road trading. The ancient city is on the Wei River about 60 miles from where it joins the Yellow River, and about 110 miles upstream from Sanmenxia Dam. Kinder made sure we knew that the infamous dam had been designed by Soviet engineers, not Chinese.

Construction of the dam began in 1954, on a 900-foot-tall plug across the Yellow River at the Three Gate Gorge narrows (not to be confused with the Three Gorges Dam on the Yangtze River). The dam was supposed to control floods and trap the river's sediments, an explicit goal being "a clear Yellow River" below the dam to halt accumulations of sediments that led to floods whenever the river jumped course. Some Chinese experts warned about the folly of purposely trapping so much sediment.

In 1963, just three years after the dam began storing water and sediment, the reservoir had accumulated four billion tons of material and lost 40 percent of its storage capacity. Hydroelectric turbines became clogged with silt. A "tail" of sediments was backing upriver, creating new flood-threats as far upstream as Xi'an. The dam's gates had to be left open and turbines pulled out so water could pass through and flush some of the accumulation. The reservoir became nearly worthless for storing water or generating electricity. In 2000, a second large dam was completed a bit further downstream on the Yellow River. Though silt-discharge channels were incorporated, that reservoir is predicted to "reach equilibrium with silts" by 2020.

The Mother River has never been easy to control.

UPDATE 2020

In January 2020, China reported modest improvements in water quality on the Yellow River, with an 8.7 percent increase in the proportion of water with better than "third class" quality, and a drop of 3.7 percent in the lowest water quality (fifth-class) status.

National Parks of Ningxia

The moment we stepped off the plane in Yinchuan, we were struck by the changed climate: though the sun was about to set, the air was noticeably warmer and drier than it had been in eastern China. We were in Ningxia Autonomous Province, about 600 miles from the east coast of this vast nation and just south of Inner Mongolia (38°30'N; 106°20'E). Annual precipitation there is only about 12 inches. Signs were written both in Chinese characters and Arabic script.

Yinchuan city sits beside the Yellow River in the shadow of the Helan Mountains. We visited several national parks in the area including Sand Lake, known for its bird life. A huge entrance arch in the shape of a crane greeted us, along with a bank of fee collection stations. One must pay the entry fee, then also buy a boat ride, the only way to really see the lake and sand dune island. We decided to side-step the steep entry fees and enjoy the birds outside in nearby fish rearing ponds, including Grey Herons, River Lapwings, Great Crested Grebes, Grey headed Lapwings, White Wagtails, Egrets, Black-necked Stilts and Hoopoe.

The same approach to infrastructure was repeated in other nearby parks; massive, gleaming entrance buildings with marble staircases and plazas, along with high entry fees plus extra charges for trams, boats, and films. It is an interesting approach to park management: landscape preservation coupled with costly "ecotourism" playgrounds.

The water at Sand Lake comes from the Yellow River. Such an abundance of surface water, there and in fish-rearing ponds that lined the highway, was a bit startling, since water-use along the river is tightly controlled by the Yellow River Conservancy Commission. Most of the diked ponds were former wetlands, converted to fish culture. China has a long history of inland aquaculture and ranks first in the world in commercial fish rearing. The major species for freshwater aquaculture are various species of native carp plus exotic Tilapia, from Africa.

The visitor center and museum at the Helan Mountains National Nature Reserve had two floors of museum displays and replica petroglyphs. We were happy to move outdoors and walk the canyon trail with the real rock art. There, hundreds of intricate petroglyphs are pecked into stone; some dating back 6,000 years.

A small creek was trickling along the bottom of the canyon. The park had informative interpretive signs and small dots were painted near some of the more faded art that might otherwise be overlooked. "Chinglish," is what our guide book called the language on signs that sometimes made for humorous translations, while still managing to communicate key messages. Along the way we read:

*"Please appreciate the sceneries and the historic sites
as you want to be appreciated"*

"Well treat the environment and nature"

"No mountaineering for hillside steep the stone loosening!"

Kinder was interested in our reactions to the signs, wanting to know if we understood the messages. Here, again, his English skills and connections with park interpreters may be able to help (though more accurate translations may take away some of the charm in those signs).

Even in that more pristine setting, park managers could not resist remodeling the creek bed into ponds lined with stones. In some of these man-made pools, large carp (goldfish) swam.

Toward sunset, we were back in the city of Yinchuan at the Yuehai Wetlands National Park. Our driver dropped us at the bottom of a huge staircase that led up to a stadium framed by Romanesque arches. Walking through an arch, we would not have been surprised to see gladiators battling on the arena below. Instead, players were kicking a ball across a soccer field at the edge of a large reservoir. The water came from a tributary of the Yellow River that runs through the city. While wedding couples were photographed amid the columns in the sunset glow, fishermen and birds enjoyed the lake. The man-made lake and its grandiose structures were built to celebrate the 50th anniversary of the Ningxia Autonomous Province. One of the lake's justifications was to restore wetland acreage lost to farm drainage in that region in the past.

The amount of total water allotted from the Yellow River to each of the nine provinces where the Yellow River passes is set by the national YRCC. In Ningxia, with 15 percent of China's irrigated farms, agriculture uses about 93 percent of the province's allotment, but that number is changing. To provide more water to cities and to coal-based industries, the government wants farm water use to drop by 22 percent. Construction of new factories and power plants can only proceed after irrigation canals serving farms are lined or repaired, to free up "new" water. By 2010, three such water-trading projects had remodeled 37 miles of canals and 105 miles of sub-streams in Ningxia to make 13 billion gallons of water a year available to new coal-burning power plants." (Ianova, 2011)

From Yinchuan, we traveled south, paralleling the Yellow River in a car powered by compressed natural gas. We were pleased by the skills of our hired drivers, but had to stay calm in the back seat because frequent passing was a pulse-raising adventure – sometimes three cars occupied the full width of two-lane roads. It seemed best to not watch. Along the river-fronting highway, we passed more channels diverting water to fish ponds, saw sheep and donkeys (on the east coast there were few grazing animals), and also an aluminum factory and a magnesium factory on the river banks, busily pouring smoke into the air.

Driving, instead of taking an overnight train, was a last-minute change of plans that allowed us to see more of the countryside and to stay a night inside Shapatou National Nature Reserve, near the city of Zhongwei. About halfway there we stopped when our GPS told us we had reached 38°00'N. A tall river-viewing tower was under construction outside the village of Qingtongxia Zhan, though not yet ready for use.

Shapatou is called "the hat of the desert," because it perches at the very edge of the Tengger Desert, where advancing sand dunes are halted by the Yellow River (37°27'N). Silk Road travelers crossed the water barrier on leather rafts. The region was short on trees for wooden boats, but was rich in livestock, so they devised an ingenious solution to the problem: they soaked sheep hides in oil and brine to make them air-tight, sewed the legs shut, and then inflated the skins like balloons. Lashed side-by-side, the hides kept wooden platforms afloat. In Shapatou National Nature Reserve, tourists ride on rafts supported by 12 to 14 skins, navigated downstream in the fast-moving current by a paddler who sometimes teaches his passengers a traditional boating song: "The Yellow River Has 99 Bends."

The meeting of river and desert at Shapatou makes a dramatic natural setting, but there, as in the other national parks, money-making "eco-tourism" activities abounded. We rode camels, the "ships of the desert," to the summit of dunes that loomed 300 feet above the river. That one-way ride deposited us near starting points for overnight camel-caravan camping trips; there were dune buggies to rent and race around on a marked course; zip-line cables spanned the river for sliding over and back; bungee jumping was available from a platform hanging out over the water and sand-sliding was a quick way down the hill.

Seeking more natural experiences, we glissaded down the steep face of the dunes, an experience that was new to Kinder (this was also his first visit to the region), and crossed a foot bridge over the river to a marsh, alive with birds. A section of the Great Wall was nearby, crumbling but still impressive. Its effectiveness as a frontier barrier must have been enhanced by the nearby river. We were 600 miles west of the more well-known sections of Great Wall we had seen outside Beijing.

Very detailed museum exhibits in the park explained desert natural history and the work carried on by the Desert Research Institute to combat the expansion of deserts. Since the Institute pioneered a system of straw thatch squares to control blowing sand, the "Shapatou Mode" has been applied

The Yellow River stops advancing sand dunes from the Taggert Desert at Shapatou National Nature Reserve in central China

to almost 400,000 acres, allowing vegetation cover to increase from less than one percent to over 42 percent in the most successful treatment areas. The effort began a half-century ago, motivated by a need to keep sand off the Bao-Lan Railway tracks that run from Inner Mongolia south through Gansu Province. Eighteen percent of China's land is threatened by desertification caused by over-cultivation, over-grazing, and logging. In China, global climate change is not a controversy to be avoided; the displays explained that climate warming is accelerating their desertification problems, with the onset of spring arriving 2.4 days earlier, since the 1980s, and a 21 percent reduction in China's glaciers in the last 50 years.

A poetic greeting was displayed in the first room of the museum:

"Here, the rolling Yellow River frames an "S" type basin,
becoming quiet tenderness...
Here, quiet and ancient houses stand with 100-meter high
hill, sand, mountain, river, park, symbiotic harmony,
thousands of years living together...
So, dear friends, forget your troubles,
hugging to the desert,
exploring the scientific truth,
feeling nature's purity;
Maintain the natural balance with your heart."

We wandered into the Desert Research Institute, which is affiliated with the China Academy of Sciences. The compound seemed abandoned until we located a graduate student, Dr. Liu, working by herself with flasks of water and soil in a laboratory. She told us she was analyzing relationships between soils and plant restoration. About 40 faculty and students use the Institute, but that day everyone else was in the field.

Despite the Institute's efforts, desertification remains a major concern in western and central China. High winds on our second day in Shapatou gave us a small taste of the sand storm potential, which can be so severe in the spring that haze from China, swept into the jet-stream, whitens the sky above our Sierra Nevada home.

Our rooms were inside the park, only 50 feet from the edge of the river, which we could see out the windows. Kinder said the hotel was "not bad for a national park." after showing us the trick for stopping the running toilet. (Toilets are not the subject of this book, but a tip for travelers to China: carry your own supply of paper; there may not be any, even in new hotels with beautiful decorative paper dispensers). In the restaurant beside the hotel a chef showed us a live fish, caught that day in the Yellow River, that he then cooked for our dinner.

Hide boats and camels at this ancient river-crossing were evidence of the Silk Road history that we would keep encountering from there across Asia. Camel caravans moved steadily between China, Europe, and India between the middle of the 1st millennium BC, with activity along Silk Road routes peaking during the Han Dynasty (200 B.C. to 220 A.D.), a surge during the Tang Dynasty (618-907 A.D.) and another while the Mongols ruled northern China (from 1220 to 1368 A.D.). From China came silk, iron, steel, bronze objects, lacquered work, and hides. Europe and India bartered glass, gold and silver objects, incense, amber, pigments, ivory, and slaves. When the Ming Dynasty sealed itself off from the outside world, near the end of the 14th century, activity ceased.

Trade was also on-going within the region itself. China required horses for defense against attack from nomads. The horses were, ironically, acquired by trading with horse-rearing nomads. During the Tang Dynasty, China paid the Uighurs one million bolts of silk for 100,000 horses each year.

According to his famous diary, Marco Polo traveled east on the southern Silk Road in 1273. The young man's father and uncle had come to the court of the fifth Mongolian ruler, Kublai Khan in 1265. After a year, they carried a message to the Pope requesting 100 missionaries be sent to Khan's court. That never happened, but two Dominican monks, plus Marco, began a return trip. The monks left the group in Syria, more interested in seeing Palestine, but Marco reached Kashgar, then Yarkand and Khotan (today's Hotan). He dictated his famous descriptions of the next 18 years spent in China to a fellow prisoner in 1298, while incarcerated during a war between Venice and Genoa.

The Taklamakan Desert was – and is – the most daunting obstacle to travelers moving across western China. The name "Taklamakan," in the Uighur language, means "the place where travelers will never get out." At Lanzhou, northern and southern routes split to circle that desert, which is about one-third the size of California. The southern branch passed through Xining then skirted the edge of the great desert along a string of oasis towns. The 38th parallel traces that southern portion of the Taklamakan, so the second half of our journey across China would take us, first, to Lanzhou and Xining, before we leapfrogged by air over the vast distance to reach a string of famous oasis towns: Hotan, Yarkand, and Kashgar.

Silk thread from silk moth cocoons in Hotan, famous as a Silk Road oasis

UPDATE 2020

"Ant Forest," a part of the Chinese multinational corporation, Alibaba, planted more than 122 million trees from 2016 through 2019 in the deserts of Ningxia Hui Autonomous region and the Inner Mongolia autonomous Region, to match "virtual trees" earned when their users' low-carbon-footprint activities saved carbon equivalent to one tree.

Up the Yellow River to Lanzhou's Green Camel Bell

The city of Lanzhou was an important Silk Road hub that became a major industrial center for modern China. With 3.2 million people, Lanzhou is the first major city on the Yellow River after it flows out of the mountains. The city is south of the 38th parallel, but just as deserts and mountain ranges funneled Silk Road caravans, major highways follow the Yellow River south before continuing westward.

Blowing sand and dust turned the sky white as we approached Lanzhou. When sand is airborne it does not appear light brown, but fills the sky with small haze particles. Major sandstorms were then occurring across northwest China, through local conditions were not so severe.

Lanzhou hugs the banks of the Yellow River for 24 miles, confined by hills on either side of the river gorge. We checked into a Super 8 hotel, the American chain, with landscape photographs of Arizona and Utah on the lobby walls and a Route 66 sign in our room. It was a short walk to the city's central square where large groups of men and women were doing a popular Mongolian dance as we came out of a supermarket. Music came from a garbage truck that drove by, as the sun went down, sounding like an ice cream vendor in the States. The song, on that warm May night, was Jingle Bells.

Liping Ran, a project coordinator with the NGO Green Camel Bell (GCB), with bright eyes and an infectious smile, set other projects aside and devoted the next two days to us. GCB's mission is to protect the environment of western China, seeking "green mountains, clear water, blue sky, man and

43

nature in harmony." The organization's unusual name was chosen, she told us, because, "Green is our dream -- where we hope to go will be green; camels are well-adapted to their environment, as are the GCB staff and volunteers, but can get lost in the vast desert, so ringing bells help people know where we are, and they bring luck."

We also met the NGO's founder, Zhao Zhong, who received a *Time* magazine "Hero of the Environment" award in 2009. Their full-time staff of six people works with two interns, two part-time staffers, and about 30 volunteers on a wide variety of projects including grass brick construction, sustainable agriculture, environmental education, and community development. The organization has an impressive list of international supporters from the United States and Europe, most importantly, since 2006, Pacific Environment, headquartered in San Francisco. On one office wall were logos for 26 organizations, including the World Environmental Fund, International Rivers, Global Greengrants Fund, World Keeper Alliance, and the Ford Foundation.

Liping began working for GCB as one of its first staffers after she received a degree in Environmental Science in 2008. "When I graduated, I thought I should do good things for the environment, so I chose this job," Liping said. "I think the youth should take responsibility for the environment," she added, using almost identical words to those spoken by the younger activist, JungRok, in South Korea.

Most of GCB's work is in Gansu Province, which extends north and south between mountain ranges and desert. The organization became very involved in reconstruction after the 2008 earthquake in Sichuan Province (southwest of Gansu), where a magnitude 8 quake killed at least 68,000 people. In Wenchuan village, near the epicenter, GCB has been supporting reconstruction with "green" techniques, using straw bale bricks instead of cement, and providing eco-toilets that need no water (helped in that project by Friends of Nature).

In the center of Gansu Province, GCB persuaded farmers who traditionally grow wheat, corn, and potatoes, to add sunflowers as a crop, because the growing plants add nutrients to the region's alkaline soil and need little water. They help farmers to form cooperatives and have been able to provide some machines, again with help from international groups. Farm family annual incomes doubled in 2009, rising by $30.

Most of the region's economy is related to the Yellow River in some way. University students are trained by GCB to be environmental educators in primary and middle schools. Their theme, in 2011, was "Yellow River history, culture, and pollution." Teams of "river protectors" are also trained to monitor chemical oxygen demand level affected by polluted wastes, collecting data they post to the national Environmental Protection Bureau's water pollution map. "We are the third audit," Liping said, "the independent party that checks these things, along with the local and national governments." Though GCB has 300 factories on an office "blacklist," only a few of those are on the national map. The *Time* article, "Heroes of the Environment, 2009," pointed out that "multi-nationals and their offshoots have taken note. After a brewery it part-owned in Tianshui ended up on the map, beer giant Carlsberg installed wastewater-treatment facilities. State-owned firms that are found polluting usually feign indifference, but at least they know they are being watched, a new and uncomfortable experience." (Ramzay, 2009)

At the time we were in Lanzhou, only 30 percent of the city's sewage was treated before going into the river, but a new sewage treatment plant was under construction. Once it begins cleaning the city's wastes, in 2012, about 90 percent will be treated. Before 2006, *all* sewage went straight to the Yellow River, along with chemical, oil, and other industrial pollutants.

"Because we are on the upper watershed of the Yellow River, we must be responsible about not polluting the river for domestic users," Liping said. She was especially proud of her successful project to improve drinking water quality in the small village of Liangjiawan (1,500 people), upriver from downtown Lanzhou. After a low "flow-through" dam was built across the Yellow River in 2007 to generate hydroelectricity, water began eddying back upstream. The nearby village had always taken its drinking water upriver

Janet (left) beside Ran Liping, a Green Camel Bell project leader, listens as women describe water quality problems in their village near Lanzhou

from the point where their untreated wastes entered the river. Now the village's intake pump began sucking water contaminated with sewage. Liping and Green Camel Bell helped organize the village leaders to seek changes and secured water purifiers for each village home, four purifying machines for the village school and a new water pipe system and purifying chemicals to serve the community.

Why would a dam be designed so poorly? China's laws do require an environmental review, but the flawed process did not identify the problem. Because the dam generates electricity with flowing water, a clean alternative to coal, the company secured carbon trade credits and impacts on local people were not addressed. We went upriver with Liping, that afternoon, to see another village with a slightly different drinking-water problem. It was nice to be outside under blue skies, with temperatures as warm as summer, Liping said. Though Lanzhou has long been notorious for severe air

46

pollution, winter is when that problem worsens, once everyone begins heating with coal.

Another small hydroelectric dam had recently been completed across the Yellow River, downstream from Cxian Shui, a small village of narrow, dirt lanes, hugging the river bank. This village depended on clean water from a community well that, two years previously, was drilled deeper to get below a layer of salty water. Liping approached several ladies in the village who seemed reluctant to talk, at first, but Liping's charm and sincerity won them over. We heard that new water pipes broke and were never repaired, so they were hauling drinking and cooking water in buckets from the Yellow River, which had to be boiled. Some people had plastic water pipes running above ground to the river. The alternative was a 10-minute walk to the well lugging heavy buckets. Once they began explaining, the women's frustration became obvious.

The young baby we met and all the people in that village deserve better. Safe and clean drinking water and sanitation is a basic human right, affirmed in a United Nations resolution in 2010. Liping's next step will be to contact village leaders to confirm the stories she heard from the residents and begin working towards a solution.

Green Camel Bell is a pioneering group in China, helping to effect change in attitudes towards the environment. Liping told us the staffers are all unmarried, completely dedicated to their work, but that it is sometimes hard to explain the value of that work to friends and family in the present Chinese culture.

The first Chinese NGO was founded in 1994. Friends of Nature focused on environmental education, at first, until the cautious central government made it clear that its role could broaden. By 2006, there were over 350,000 NGOs officially registered with the Ministry of Civil Affairs. With rapid economic growth, enforcing compliance with national environmental regulations had become difficult. The NGOs serve as that useful "third audit," described by Liping.

On our second day in Lanzhou, Kinder departed for Beijing to meet another group. With Liping, we went to see 60-foot tall waterwheels, ingeniously powered by the flowing current of the Yellow River to lift water high above the river banks. Similar

In Lanzhou, hundreds of wheels lifted Yellow River water to irrigate farmland and provide water to the city

mills had served this city, local farms, and grinding mills ever since the Ming Dynasty (1368-1644 AD). In 1952, Lanzhou,"the waterwheel city," still had 252 of the big wheels lining both banks of the river. Some of those still operating are hundreds of years old.

Massive logs in the waterwheel frames were evidence that the region once had big trees growing on forest hillsides that today are mostly treeless. Beside the highway coming into Lanzhou, terraces were planted with small conifers, part of the national tree planting campaign. Japanese groups have even come to Lanzhou to plant some of those trees, because the spring dust storms blow their direction.

Liping arranged a meeting at a teahouse with one of her university professors, Zhang Songlin, who teaches geography and environmental science classes at Northwest Normal University. She translated as the professor told us about sustainable development issues in China. "All countries need to have sustainable development," he said, "but many of our local governments focus on higher GDP." Throughout China, whenever the subject was economic growth and development, they used this shorthand, "GDP," for gross domestic product.

"Officials, while serving their terms in office, need good results so they can go out at a high rank. People need jobs. Young people need an income to buy a house, get married, and become self-sufficient. But the old Chinese culture respected nature. People and nature were thought of as one. In old times, the belief was that if you harm the environment, you would be punished. Now, increased GDP is the belief. We don't trust each other enough. We can't trust food safety. Air quality is improving little by little, and also water quality, but I am not very hopeful about the future." Despite that pessimism, Professor Songlin agreed with us that Green Camel Bell's work was a reason for hope.

That night, Liping escorted us to the train station. Her next task was in Wuwei, a city right on the 38th parallel, to begin another water quality investigation.

UPDATE 2020: GCB has added a climate change focus on low-carbon energy efficiency, organic agriculture and farmer cooperatives, and continued protection of the "Kidney of the Yellow River" wetlands.

Qinghai, Blue Lake of the Tibetan Plateau

Xining is the home of the Qinghai (pronounced "ching-high") Salt Lake Institute. In 2008, we had seen a news story in the Chinese press headlined *"Massive Program Launched to Save Qinghai Lake."*(Xinhua news Agency, 2008). That parallel to the historic efforts to save Mono Lake, plus the fact that China's largest saline lake occupies a watershed intersecting the 38th parallel, made Qinghai Lake, east of Xining, a must-see destination.

A researcher at Mono Lake wrote to introduce us to Professor Fafu Li from the Qinghai Salt Lake Institute, who was our gracious host for several days. Fafu arranged a meeting with Institute scientists and graduate students interested in hearing about Mono Lake and our 38th parallel project. He had worked for 19 years at the Institute.

One of the faculty members at our meeting was Dr. Junqing Yu, the author of a book (written in English) about Qinghai Lake. He particularly wanted to compare the chemistry of Mono with the vast inland sea of central China. Qinghai Lake is saline, though less salty than Mono Lake or the ocean (at 14 grams per liter dissolved salts). Its water is quite alkaline, though, with a pH of 9.2 (Mono Lake water is pH 10, similar to mild detergent). Qinghai is super-saturated with dissolved aragonite and calcite, the two common crystal forms of calcium carbonate and the main ingredients in Mono Lake's tufa towers.

Lakes like Qinghai and Mono that occupy closed basins with no outlet gradually accumulate salts dissolved from the watershed by tributary streams. The Earth's oceans are salty

50

for the same reason. Qinghai Lake's watershed valley, drained by six rivers, covers more than 11,400 square miles. Most of its water arrives via the Buh River which extends northwest almost 200 miles to the crest of the Datong mountains, just north of the 38th parallel. The Qinghai Nanshan mountains, south of the lake, approach the shore more closely. As we saw during two days of exploring, the Nanshan provide a scenic snow-covered backdrop, something like the mountain setting at Mono Lake. Qinghai's water surface covers 1,700 square miles – a huge inland sea vastly bigger than Mono Lake, more comparable in size to Utah's Great Salt Lake. It becomes very obvious on maps, as it is the only large water body in central China.

Fafu and a driver took us to the lake, pleased, it seemed, to be escaping from the office on an overnight road trip. From Xining, we crossed a pass at 11,250 feet above sea level and left the Yellow River watershed behind. The surface of Qinghai Lake is 10,154 feet above sea level. It took most of the day to travel around the gigantic lake to its west shore, 220 miles from Xining. Grazing animals abound in the grasslands along the lake, and we got our first close look at yaks *(Bos grunniens)* – thousands of yaks and even more sheep. The diet of the local Tibetans is almost entirely meat, especially mutton.

Buddhist prayer flagging was strung in many places around the basin. The Mongolian name for the lake, "Kokonor," means "blue lake." They, and the Tibetans, consider it a sacred place. Here on the shore of the vast sea, a Mongolian leader conferred the title "Dalai Lama," for the first time, in 1578, to a spiritual leader named Sonam Gyatso. "Dalai" is Mongolian for "ocean, while "lama" is a Tibetan word for priest.

The main tourist destination at Qinghai Lake is Niao Dao National Nature Reserve, also called simply "Bird Island," on the west shore. Each year, 189 species totaling about 100,000 birds migrate through or nest there. Bar-headed Geese *(Anser indicus)* arrive in the spring after crossing the Himalayas from India at elevations over 25,000 feet.

After the long car trip, we were eager to walk, and enjoyed the 200 yard stroll through an underground tunnel, wide and well-lit with photographs of wildlife and scenery, that leads to a viewing blind where nesting geese were just a few yards away

51

on the other side of glass windows. Thanks to the lengthy approach that hides people out-of-sight, the birds were unaware that dozens of people were within a few yards, watching their intricate courtship displays and flights. Here, the Chinese inclination for grand structures had produced the most spectacular bird blind we had ever seen.

An electric shuttle bus dropped us at another observation point above a nesting colony of Common Cormorants (*Phalacrocorax carbo*) and Black-headed Gulls (*Larus ridibunus*). A trail led down to the edge of the lake. The water tasted slightly bitter, and we smiled at the familiar sight of suds forming where waves agitated the water on the shore. The lake supports diatoms and various kinds of algae that feed just one kind of fish, the endemic Scale-less or Naked Carp (*Osteichthyes cypriniformes*).

The carp never showed themselves. They have been protected from fishing for almost 20 years, yet, at the hotel restaurant that night, our waitress asked Fafu if we wanted to order the fish! He declined and told us that it would be "very illegal;" he could lose his job if he had said yes. It seemed a "fishy" situation, that the protected carp was available there on a menu that otherwise featured meat and potatoes.

Grassland surrounds Qinghai Lake in an almost treeless landscape, where sheep, goats, and yaks graze, competing with diminishing numbers of wild Przewalski's Gazelles (*Procapra przewalskii*), an endangered species. Much of the grassland close to the lake has been converted to rapeseed *(Brassica napsus)* farming. Publicity photographs for Qinghai Lake always feature the bright yellow flowers that bloom in the summer. The plant is valued for making canola oil, plus by-products of oil production that include animal fodder and "oil-cake" fertilizer. Fields were brown and stubble was burning when we were there; it was still early spring at that altitude.

Rapeseed farming requires irrigation that contributed to the decline of the lake, which lost 10 feet of its depth and shrank by about 150 square miles between 1959 and 2006. The government's 10-year plan to "save Qinghai Lake" includes actions across the watershed to maintain the lake level and improve ecological conditions. Some farm fields have been taken out of production; sheep numbers are to be reduced by

Black-headed Gulls at Qinghai Lake, China's largest alkaline lake

about 1 million animals, and 2 million acres of pasture are to revert to grassland. "Ecological migration" of the region's nomadic herders will relocate 4,000 people.

To reduce livestock numbers, individual, fenced grazing parcels have been allotted to the traditionally nomadic people. In several places along the southwestern and north shores we saw new houses provided by the government. The nomads are still allowed to roam with their herds to higher ground during the summer, but total numbers of animals are controlled. Additional justifications for the housing change, says the government, are clean drinking water, improved housing conditions, better diets, plus financial support for local basketry and handcraft production.

The government has also been poisoning rodents, which they say degrade grasslands and compete for forage with livestock. That program is controversial. Some researchers consider the targeted Plateau Pika (*Ochotona curzoniae*), a "keystone species" on the grassland ecosystem of the Tibetan Plateau, providing food for predators, and burrows to birds seeking breeding shelters. The only pika we saw were mounted specimens in a museum.

On the north shore, Kangtsa is the largest town in the lake basin. It is a farming center and the capital city of a

Qinghai Lake's alkaline water generates suds when waves lap the shore

county with the same name. Nearby, a railroad track that connects the area to Lhasa was being converted for electric locomotives. New electric lines were running everywhere, some from solar and wind generating plants. The high plateau must have looked very different just 20 years ago. The region illustrates a dilemma facing central and western China, where public services are improving, but at a cost to traditional ways of life.

After circling the north side of the lake on a very rough road, we stopped at Qinghai Lake National Park. Here there had been a grand plan for lake cruises. Docks and other infrastructure were built, but the government decided that boat engines were too polluting and halted the project. In fact, no boats of any kind are allowed. Though the pot-holed access road and abandoned facilities gave the park a ghost-town feeling, there was a bus group of tourists there that day and a crew was stabilizing sand dunes using the straw squares developed at Shapatou.

The elevation, clean air, and endless vistas reminded us of home.

Before leaving Qinghai Lake, we crossed a 12,500-foot highway pass into a separate basin and explored hypersaline Chaka Lake. Only bacteria live in its concentrated brine, as in the Dead Sea of Israel. Salt works have been here for centuries. Fafu explained that Chaka could provide enough domestic salt for the entire world for 500 years! Large barges moved out a channel past evaporation ponds toward the middle of the lake. On foot, far from the open water, we walked along a jetty lined by a layer of salt a foot thick.

While lunching at a small noodle cafe, mostly serving locals, the waitress suddenly told us to hurry and finish because a serious sandstorm was blowing in. We barely beat it back to Xining.

Under skies turned white by blowing sand, we flew across the Taklamakan Desert to Urumqi in the Xinjian Uighur Autonomous Region.

UPDATE 2020

In March 2020, the provincial government reported that Qinghai Lake had continued to rise with improved water flows back to the lake, bird species in the area had increased from 164 in 1996 to 225, and the lake basin's unique naked carp fish stock increased from 2,857 tons in 2002 to 102,515 tons in 2019.

Hotan to Kashgar on the Southern Silk Road

The tourist map for Xinjian states that Urumqi, the transportation hub in northwest China, is the farthest city from the ocean in the world, over 2,000 miles from Beijing and the east coast of China. Our next goal was actually Hotan, one of the oasis towns along the Southern Silk Road to Kashgar. Resolving some confusion about the Chinese spelling of that city as "Hetian" on airport signs, we made our connection and were happy to see our Uighur guide, Abdul Wayit, and a friend from home, Rick Kattelmann, waiting for us at the Hotan airport. Rick is a retired hydrologist, now a serious photographer, who had traveled here before and recommended Abdul's guiding services. Abdul, 23-years old, handsome and ambitious, had learned English and become a guide as a way off the family farm. He was eager to share his Uighur culture and became an invaluable window into the soul of his people.

Hotan is famous as an oasis on the Silk Road and especially for its jade and silk. The sprawling settlement is an island of green in the Taklamakan Desert, with orchards of fruits and nuts and fields of wheat going on for miles. At the eastern edge of town, men were planting and drip-irrigating peach trees and tamarisk as part of the "green wall" that holds back the shifting dunes.

Hotan exists because of its water – the White Jade (Yurungkax) and Black Jade (Karakash) Rivers, which flow from the Kunlun Mountains. After the two merge north of the city, they become the Hotan River, which continues north (and soon dries up, except in the wettest seasons) into the Takla-

At the edge of the Hotan oasis, men planted peach trees to keep the Taklamakan Desert at bay

makan Desert. Hotan traded in jade with Mesopotamia from the third millennium onward. A new discovery in 2001 in the river bed gravels started a modern day "jade rush," a frenzied free-for-all that has torn up the riverbed using everything from shovels to huge excavators. People sleep among the tailing piles and work with lights at night, looking for the shine of the precious stones. In an informal jade market near the river (a parking lot gathering) men haggled over stones. One jade buyer showed us a golf-ball-sized piece that, he said, was worth 5000 Yuan (about $1,600). He glanced at a small rock we found and tossed the worthless river stone contemptuously on the ground.

The jade found here is rarely green. Most often it is "mutton-fat jade," opaque, creamy white, with tan or orange tinges. Abdul told us that many people in Hotan had become wealthy by prospecting, as the market price per ounce is as much as 40 times higher than gold. "Eighty percent of the people in Hotan now have cars," he said, attributing that to jade trading. Local Uighurs sell their finds to Chinese middlemen who serve the demand from growing numbers of wealthy Chinese. In Chinese culture, jade is the most valuable of precious stones, symbolizing perfection, constancy, and immortality.

Those ethereal qualities contrast with the toil of thousands of miners and hundreds of large earth-moving vehicles that we saw turning over the river bed. The channel looked like the aftermath of the California Gold Rush, where mounds of dredged gravel are still seen, 150 years later, heaped along the banks of foothill rivers.

When melting snow sends high flows down toward Hotan, the torn up channel can no longer confine the water effectively, so floods have become a threat (as they did in California by 1860). In 2007 the Chinese government issued a notice to stop the heavy equipment work, but prospectors ignored the order, which was not enforced by local officials. We saw plenty of excavators at work, along with small groups of people wielding shovels.

Silk, rather than jade, has been Hotan's most famous and enduring trade item for centuries. A factory in Hotan still does the whole process by hand, extracting the thread from the cocoons of silk moths, then spinning, dying and weaving fabrics for garments and tapestries.

Farming has been the traditional support for most of the population of Hotan, which is renowned for its almonds and walnuts, plus fruit and vegetable staples. Thanks to irrigation, the greenery of the oasis city was more extensive than we expected, though it was just a short drive from the center of town to the edge of the desert.

Walking across dunes gave us a feel for the abrupt change. A half-mile into the desert, a tomb and small mosque marked the burial site of Imam Asil, who died about 900 years ago, during a war between Uighur Muslims and Tibetan Buddhists that went on for hundreds of years. Pitchers with water for washing before prayer were being filled from a well. A bucket is lowered and raised dozens of times a day. We took turns at the crank, confirming something we already appreciated: water is heavy; eight pounds per gallon, hauled up 20 feet at a time.

That night in Hotan, Abdul asked if we wanted to "eat the food Marco Polo took back to Italy?" The spaghetti was delicious, though Marco Polo did not eat his with tomato sauce in the 13th century. The legend of Polo introducing pasta to

Italy is a popular story in that region, but Italians actually had been making pasta for centuries before Polo came to China.

Driving west from Hotan, we stopped in the village of Kargilik to explore a colorful weekly market. Underneath red tents, the colors of silk clothing and textiles glowed with fiery brilliance.

Even with an air-conditioned van and a good driver, the 350-mile, often bumpy, trip to Kashgar felt long. What must it have been like to do this same trip on a slow-moving camel? A railroad is set to open soon along the route, and a new highway was under construction. The evolution of the Silk Road continues.

It was a relief to break up the trip with a night in Yarkand (a "small" town by Chinese standards, with only 560,000 people (38°17'N). On a walk through the oldest section of the city, we asked Abdul about a sign painted on a mud wall. "We only have one Earth, so practice family planning," he translated, adding that the government had placed the sign there.

"We agree with that message very much," we told him. "What do you think?"

"I think it is good," Abdul said, then grinned and added, "but I still want five children." Large families are a way for Uighurs to fight back against their feeling of invasion by the Han Chinese that keep moving to the western province. Though the government limits rural Uighurs to three children, it is apparently possible to circumvent the limit if babies are born in different localities.

Northwest of Yarkand, the desert suddenly turned green again when we passed a reservoir outside Yengisar, "the town of knives." Abdul asked if we wanted to stop and shop for a knife, but of course there was no way to bring one home in our carry-on luggage. He told us that Uighur men used to always carry knives on their right hips, adding, "Our knife is our manhood." But after violent demonstrations in recent years, men are no longer allowed to openly carry a knife. "If I wore one now, soon I would belong to the police," Abdul said.

The Uighur region seems like a whole different country than the rest of China. For much of human history, it was. Veiled women mix in the business districts with Han Chinese

At the Sunday animal market in Kashgar, sheep, goats, cows, donkeys, or fast horses for racing are for sale. Note the money held by the man making an offer to the rider

in mini-skirts and high heels, but the two cultures segregate into residential areas of the cities. After China took over in 1948, the province was named Xinjiang, meaning "New Frontier." The Han living in China's crowded east are encouraged to move west, an influx generating tension that has erupted into violence in Urumqi and Kashgar. Just a year before our trip, our friend Rick had toured the same region and seen, everywhere, a heavy presence of military and police. Happily, things had calmed down since then (but a year after our visit, new violence flared and the police shot several demonstrators).

Along the highway, a truck was parked in our lane while the driver negotiated with a sheepherder for some of his animals bunched beside the highway. What sheep found to eat in that desert was a mystery; it appeared nearly barren (in part because there were so many sheep and goats nibbling whatever grew). The purchased sheep were likely going to the Sunday animal market in Kashgar.

Kashgar is famous as a Silk Road trading center and crossroads, bridging China and Central Asia. Pomegranates and apricots are the specialty crops grown near the city with river water running off the Karakorum and Kunlun mountain ranges. Marco Polo wrote in the 13th century that the "inhabitants have very fine orchards and vineyards and flourishing estates. Cotton grows here in plenty, besides flax and hemp. The soil is fruitful and productive of all the means of life. This country is the starting-point from which many merchants set out to market their wares all over the world." (Polo, 1982, 65)

Outside Kashgar, we visited Abdul's family farm where he was born and raised, and we met his mother, father, and wife. It was a great honor to be invited into a Uighur home. Knowing they were a conservative culture, Janet asked if she should cover her head with a scarf while in their house. "No," Abdul answered, "they understand you are not from here."

The farm grows corn, wheat and apricots (the latter are often eaten green), and they raise a few cows, goats, chickens, plus a friendly cat. Adbul's wife cut and stretched noodles, her flashing fingers working the dough before she cooked the noodles. We gave his mother a Mono Lake
shopping bag and she seemed thrilled after she saw how it stuffed into a small sack.

The famed markets of Kashgar are incredibly colorful, with dramatically dyed fabrics, luscious fruits, and shiny copper pots. Women go to market in exquisite dresses that, Janet said, she would only consider wearing to a formal wedding. Livestock are brought each Sunday to a special market, where sheep, donkeys, cows, goats and even cats and dogs have been sold or traded for centuries. There was an area to "test ride" the sale horses at top speed.

On the way to the market, we drove by Kashgar's cellphone street market. In the 21st century, people buy and barter for items that have been manufactured for thousands of years, while carrying on cell phone conservations.

Many buildings in Kashgar's old town were being torn down and rebuilt with fired bricks and reinforcing steel

Silk shawls and blankets are sold in the bazaars of the oasis towns along the southern Silk Road

instead of mud bricks. The government says the changes are necessary to protect against earthquakes, but Abdul feels the old city withstood many quakes with no problems. Forced renovation contributes to a perception that the Chinese Han are undermining the Uighur culture. The government insists it respects minority cultures and will keep most of the old town intact as a tourist attraction – with an admission charge.

Walking along Kashgar's streets, as in every city in China, can be a terrifying adventure. Chinese drivers do not slow down for pedestrians. One learns to follow the "herd" at crossings, to wait on the line in the middle of the highway as cars speed by too close for comfort, then proceed when it is safe. Traffic was less hectic in the old town, where we saw abalone shells for sale, though we were thousands of miles from the ocean. The shiny material is used to ornament items like knife handles. After watching a man skillfully turning wood on a lathe, we bought several small items, keepsakes from that oasis trading center on the 21st century Silk Road.

UPDATE 2020

Since 2017, China has forced over a million Muslim Uighurs in Xinjiang province into indoctrination and coerced labor camps. We are concerned for Abdul Wayit and his family, but do not know how they are faring.

The Edge of China and Turkmenistan

The Karakorum Highway (locals call it "the KKH") connects China to Pakistan on the most westerly of the highway passes across the Himalayas. We were eager to compare the high country along China's western border to our home in the Sierra Nevada.

A cutoff to the north from the main highway led to Oytagh Glacier, at the head of a scenic pastoral valley. "Oytagh" means "mountain basin." The valley community was raising crops along the river, backed by towering mountains, while livestock grazed on the new spring grass. Yaks, camels, and Kyrgyz yurts, hats and clothing appeared here, close to the border with Kyrgyzstan. It felt good to breathe clear mountain air. The river was muddy with water because the glacier, hanging below peaks towering 21,000 feet, was melting fast.

The Oytagh Glacier viewpoint is 9,533 feet above sea level. Just as we reached the viewpoint, an avalanche dropped from the cliff west of the distant glacier. To the left, a waterfall plummeted over the edge and fell as straight as upper Yosemite Falls for hundreds of feet. Or farther, perhaps. Scale was hard to judge in the grand landscape.

The view from the end of the trail was once straight down onto the lip of the glacier, but the ice had retreated drastically and now was at least a half-mile away, leaving behind only frozen humps covered with black silt. Glacial retreat there was even more dramatic and obvious than in the Sierra Nevada mountains of California. Some of the clearest signs of atmospheric warming can be seen at the world's higher elevations. Rapid glacier melting in the mountains of northwestern China will eventually become a water-supply concern for people downhill, as glaciers have been the dependable water source for rivers that serve the region's farms and cities.

Though we had planned to sleep that night in a yurt, we learned they were only available to Chinese tourists, not foreigners. Water was shut off inside the park, including the restroom buildings serving the yurts, but also in the hotel built for foreigners, due to damage from frozen pipes that winter. Big jugs of water were provided for toilet flushing. The restaurant next door was also not quite ready for business. They offered to kill a sheep and barbecue kebobs, but we were too hungry to wait that long, so made do with our own supplies. A thermos of hot water was left outside the door in the morning so we could make instant coffee.

Next morning, in the valley village, kids gathered in their schoolyard to salute the flag while a recorded anthem played. A local bus was taking on passengers, including two live sheep being settled into the rear luggage compartment where suitcases are carried on urban routes.

The Karakorum Highway continued its climb up toward the Pamir Mountains. A herd of camels grazed in a wash and, after photographing them, we asked Abdul if they were running wild. He said, "There are no wild things in China." (Happily, that is an overstatement; we had seen lots of wild birds, though few wild mammals). The road also passed construction sites in Gez River canyon for a hydroelectric generating plant and, at the head of the canyon, a new dam. An 800-foot drop from the dam would turn turbines to generate 200 megawatts of electricity. Most of the construction workers had been trained while working on the 3 Gorges Dam on the Yangtze River.

Sand Dune Lake (not to be confused with Sand Lake near Yinchuan), will be elevated by the dam. We stopped to photograph yaks and yurts and immediately attracted men selling malachite and jade jewelry. There, at 10,770 feet, the road turned directly south toward Karakul Lake. In that remote location surrounded by mountain peaks, Abdul's cell phone rang. He seemed surprised that we were amazed by the phone reception; China has exceptional coverage. The call was from our Chinese guide, Kinder, checking to see how we were progressing.

Yaks and cattle graze beside Karakul Lake, near Muztagh Ata mountain

Travel guide books name Karakul Lake, 11,999 feet above sea level, the most beautiful lake in China. Clouds obscured the nearby mountain peaks that afternoon, but when we returned on the following day, there was a clear view of snow-covered Muztagh Ata, the 24,757-foot mountain that towers over the south end of the lake. To the east, clouds continued to hover around two Konggur Peaks, both higher than 25,000 feet. Kyrgyz nomads live near the lake, raising sheep, goats, horses, donkeys, yaks, and camels.

Janet went for another camel ride along the lake shore. Those amazing creatures can handle everything from hot desert sands to cold, rocky mountain trails. Seeing them was a reminder that here, high on the Pamir Plateau, we were still on the Silk Road.

Kyrgyz people live in yurts and raise yaks in western China

We drove up to 13,374-foot Tashkurgan Pass, and passed the "Welcome to Tashkurgan Tajik Autonomous County" sign, written in Chinese and English. Here, at 74°55'E longitude, was our point farthest west in China, 44 degrees of longitude from the Yellow River delta.

The highway plummeted almost 3,000 feet to an intersection where trucks were turning west toward the international border with Tajikistan, only seven miles away. We continued on the KKH, watching our GPS for exactly

38°00'N, and stopped there for photos in front of a small mosque, not far from Takiman village.

About 70 miles south, the highway would crest again at Kunjerab Pass (15,397 feet) and the border with Pakistan. We stopped, instead, at Tashkurgan, just as Chinese caravans had done for centuries to transfer their wares to Central Asian traders (37°47N). Beyond there, branches of the Old Silk Road led south to Kashmir and India. The westward route continued through Kabul (Afghanistan), then south of the Caspian Sea via Tehran (in Persia, now Iran), before turning back toward the 38th parallel and Mediterranean ports in Turkey.

Tashkurgan was our final destination in China. Fortress ruins on the hill above the city date to the 8th century, 500 years before Marco Polo passed through. The culture of neighboring Tajikistan dominates the Tashkurgan population, so traditional dress was again different, with women wearing pillbox hats covered by scarfs.

Crossing China was fascinating, infuriating, mysterious, and wonderful. We took away memories of vast empty deserts and mountains, wild places and mostly wild park-lands that persist along with crowded cities and the world's largest population. Still a Communist nation, officially, capitalist enterprises burgeon across 21st century China and the central government does not have total control. It was a pleasant surprise to learn that NGO groups like Green Camel Bell are welcomed by the national government because they implement and help enforce environmental laws in distant provinces – a sign that China is not simply ignoring the tension between rapid development and its unwelcome side-effects.

UPDATE 2020

The Hindu Kush Himalaya Assessment 2019 found that even if carbon emissions are dramatically cut to limit global warming to 1.5 degrees C, over a third of the glaciers in the Hindu Kush Himalaya region will have vanished by the year 2100. If emissions are not cut, this glacial loss could increase to two-thirds, with enormous consequences for those relying on that water.

Returning to Urumqi to fly west on the 38th parallel, we passed above desperately poor Tajikistan, war torn northern Afghanistan, and a "finger" of southern Uzbekistan to reach Ashgabat, the capital of Turkmenistan.

When the GPS screen showed we had reached the 61°East longitude in Turkmenistan, we stopped the car. Back home, at Mono Lake, the 38° latitude line intersects the 119°West longitude; 61 + 119 = 180°, so we were halfway around the globe in Turkmenistan, where local clocks were 12 hours different from California.

In Ashgabat, the capital and largest city (37°58'N), we had been met by our next local guide, Berkeli Atayev, tall, dark, and welcoming. Berkeli had visited Great Salt Lake seven years earlier with a group of Turkmenistan Tourism officials. He drove us straight east for five hours toward Merv, the most influential ancient city they never taught us about in World History class.

Half-way to Merv we reached that half-way-around-the-world longitude point beside a reservoir that stored Karakum Canal water. The canal dates back to 1954 when Turkmenistan was part of the Soviet Union, and it diverts 50 percent of the Amu Darya River, a major tributary of the Aral Sea. It supplies 90 percent of the nation's water, moving it 540 miles to Ashgabat and to farm lands along the route, making it the second longest aqueduct in the world. (San Diego reaches 444 miles via the California State Water Project; the Los Angeles Aqueduct extends 338 miles into the Mono Lake basin; the city of Sirt, in Libya, moves water the farthest – 745 miles from sub-Saharan aquifers to the north coast.) (Shea, 2011). With the water, Turkmenistan grows cotton, a crop farmed there for 2,000 years, and also fruits, vegetables, rice, and wheat on the tiny portion of the country (four percent) that is arable. Most of the nation is covered by the Karakum Desert.

Because of diversions by Turkmenistan and other nations, the Aral Sea, the fourth largest lake in the world before the 1960s, lost 90 percent of its volume. Its water level dropped more than 60 feet, causing the sea to split into two small lakes. Salinity increased to lethal levels for fish and a vast acreage of toxic salt-covered land was exposed to the wind. In Uzbekistan and Kazakhstan, downwind, respiratory diseases and throat

cancer rates soared.   The Aral Sea's problems originate in
other nations and, so far, little has been done to address one
of the world's greatest environmental tragedies, though a dam
built to separate the remnant in Kazakhstan has improved
conditions there, on a small scale, for fish, birds, and
residents.   In 2010, Turkmenistan hosted a meeting in
Ashgabat with Uzbekistan, Kazakhstan, Kyrgyzstan, and
Tajikistan, the four central Asian nations with the most stake
in the Aral Sea.   Despite exchanging high-level views and
issuing statements about the importance of solving the Aral
Sea's problems, effective action has not yet followed.

Though the Karakum canal spread agriculture across
southern Turkmenistan and addressed the thirst of its cities,
the unlined canal has also leaked water along its route for a
half century, dissolving salts in the ground that are then
drawn to the surface by evaporation. Ironically, the irrigation
canal increased salinity to levels unsuitable for crops on two
million acres and fouled local wells near the canal.

The government hopes salinity problems on southern
farmlands will be improved by sending 2.6 trillion gallons of
salty irrigation runoff water each year to a natural depression
in northern Turkmenistan, creating *Altyn Asyr* (Golden Age)
Lake. Begun in 2000 (to be complete by 2020), a 1,650-mile
network of canals will fill the basin until the lake reaches a
surface area of 770 square-miles and is 230 feet deep.   The
quality of the water in the resulting lake (perhaps better
thought of as a sump) is a concern. Other claimed benefits of
Turkmenistan's new man-made lake are as habitat for
migratory birds and as a new source of agricultural water that
will make the nearby desert lands bloom with plants and
livestock.

The predicted benefits may be short-lived. Life tends to
follow water, certainly, but how much transformation can
happen in the desert will be limited by the salinity and toxicity
of the runoff water. The Salton Sea, in southern
California, also serves as a low elevation collection point for
runoff from Imperial Valley farms, but accumulated salts (plus
fertilizers and pesticides) are now approaching concentrations
lethal to fish, with impacts passed along to millions of
migratory birds that depend on the lake.

The 4.5 billion dollars (in U.S. currency) spent on the Golden Age Lake project could instead be used to finally line the Karakum Canal and return the conserved water to the Aral Sea, but this lake project has been a favorite scheme of Turkmenistan's last two presidents (dictators-for-life) who

A few mud-walled fortresses and buildings still stand in ancient Merv, an ancient Silk Road city in Turkmenistan

both have embraced many grandiose Soviet-style construction projects.

Our goal that first day in the country was the city of Mary, which sits next to (and in some places sprawls into) ancient Merv, a State Historic and Cultural Park. Our guide was Yevgenia Golubeva, of Russian heritage, assigned by the national tourist bureau. Merv was one of the most important cities along the Silk Roads of Central Asia, an administrative, trading, military and religious center. Different cultures took turns dominating the city, each building its own mud-brick walls, houses, fortresses, and mosques. Greeks were in charge in the first few centuries B.C., backed up by Alexander the Great's army. Merv may have been the largest city in the world from 1145 to 1153, under the Seljuk Turks, with a population of 200,000. (Rosenburg, 1987). The Persian mathematician,

Omar Khayyam, used an observatory there; he is perhaps most famous in our country for his poems published as *The Rubaiyat.*

Merv existed at that location because of water brought from a reservoir on the Merghab River. The Merghab flows from the mountains of Afghanistan and creates a lush valley, a welcoming contrast for settlers and travelers to the harsh Karakum Desert. Ironically, the water that made the city possible was also its Achilles's heel, taken advantage of by sons of Genghis Khan. They arrived with an army of 8,000 soldiers in 1221 A.D., were repulsed by defenders at the walls, but finally forced surrender by destroying the city's dam on the river. With no water supply, gates were opened to the Mongols, who proceeded to massacre its inhabitants. Merv's ancient glory days came to an end.

Our tour, on a hot, sunny day, included the forts and mosques still standing, though much of the ancient city's mud structures have dissolved away or been destroyed. We found cool shade inside a 12th century ice house, though its roof was missing; double-walls still trapped an insulating dead-air space, like a thermos, as nesting swallows flew in and out. Camels grazed in the historic park, too, though Turkmenistan had single-humped Dromedaries, instead of the Bactrian two-humpers of western China.

On our return to Ashgabat, we visited downtown markets and city museums. We toured a stable for some of the nation's fabulous Akhal Teke horses, a type that may have been the source for the Arabian breed. The national museum's exhibit about the importance of horses in the Turkmen culture called them "the wings of a young man," and quoted a local proverb: "Getting up early in the morning, first greet your father, then greet your horse." A new horse-racing arena was adorned with statues and gleamed with marble and gilt trimming. Elegant stables housed the President's retired horses.

Massive architecture is the rule for government buildings in Ashgabat, contrasting with Soviet-era cinder block apartments on nearby streets. The image of Turkmenistan's president appears in all public places. Ashgabat felt like some combination from "The Wizard of Oz" and "Dr. Strangelove."

No other tourists were on the streets. Berkeli would like people from other countries to know that, "We are kind, hospitable people and would like to see more tourists. Though our economy is mostly based on minerals and cotton, there are lots of possibilities for tourism, which would provide people with jobs and an income. For example, Turkey's GDP is based 20 percent on tourism, but Turkmenistan's is only a fraction of 1 percent."

Unfortunately, authorities controlling travel into the country are not so welcoming. Paperwork for tourists is daunting, including a required "invitation" to enter the country before securing a visa and a sheaf of travel papers that must be carried everywhere. The bureaucracy seems to distrust foreign travelers. At the Ashgabat airport, before leaving, our bags were scanned at four different locations and our passports checked six times before we reached the waiting room.

The 38th parallel extends from Turkmenistan's western border across the south end of the brackish Caspian Sea. We flew above that great landlocked lake, the largest in the world, famed as a source of caviar. Seven species and sub-species of sturgeon provided eggs for the delicacy, but the fish have declined to the edge of extinction due to overfishing and water pollution (pesticides, heavy metals and oil), and because dams block spawning access up the Volga and Kura rivers. Beyond the Caspian Sea, our flight passed over a small arm of northern Iran to reach eastern Turkey.

UPDATE 2020

The year 2018 brought bread-rationing in Turkmenistan, with long flour queues. The nation's economy had depended most on petroleum and natural gas, and prices plummeted in 2015-2016. In May 2018 the U.S. banned imports of cotton good from the country, where child labor and forced labor for cotton harvesting occur. No effective steps have been taken about the Aral Sea decline, and in February and July 2019, massive dust storms layered alkaline dust from the exposed Aral bed onto croplands in northern and eastern Turkmenistan.

Hasankeyf in Peril and the Tigris and Euphrates Rivers

A vast inland sea with alkaline water whose strange chemistry produces large calcium carbonate structures underwater, in a dramatic setting with snow-covered mountain peaks, and a lake surface that is thousands of feet above sea level...all of that sounds exactly like Mono Lake, our home on the 38th parallel. It also describes Lake Van, the largest lake in Turkey. Van was our first stop on a 15-day trip across the width of this very large country.

Lake Van also differs from Mono Lake in significant ways. It is much larger, with a 270-mile shoreline (Mono's is about 40 miles); an average depth of 560 feet (today, Mono Lake averages about 60 feet deep), and with a deepest point that is 1,480 feet (the deepest place in Mono Lake is only about 150 feet).

Van holds a lot of water, but not so much salt. It is the largest soda lake on Earth, dominated by carbonates, also containing sulfates (both key ingredients in Mono), but without our home lake's chlorides – the ion associated with dissolved table salt. There are fish in Lake Van, though only a single endemic species, the Pearl Mullet (*Chalcalburnus tarichi*) adapted to the unusual water chemistry; Mono Lake is far too salty for fish. The alkalinity of both lakes is fairly close, around pH 9.8 for Van and 10 for Mono. Because Van's water is less harsh, it does support the mullets, along with hundreds of species of plankton and many migratory and nesting birds. Lake Van even has calcium carbonate structures chemically related to Mono Lake's tufa formations.

Van's are out of sight in the depths, where cyanobacteria build aragonite towers up to 130 feet tall.

We arrived in the city of Van on the southeast shore of the lake (38°29'N). Adem, a student about to graduate from medical school, was on our bus, and as many other helpful Turkish people would do, he introduced himself, answered all our questions, was interested in our project, and he even walked with us through the city to be sure we found our hotel. Following his directions, we took a mini-bus out to the lake shore where local kids were swimming and local women and their babies enjoyed the sun on a dock. We felt the water, which was slippery, and our wet fingertips tasted like baking soda. The water here, as at Mono Lake, can be used to wash clothes. Salts extracted from the lake are sold as detergents. Climate scientists drill down through Lake Van's sediments to examine a half-million years of climate data.

Women and children on a Lake Van pier in far eastern Turkey

There are refugee camps near Van, where desperate Kurds are escaping political problems in neighboring Iran. Far eastern Turkey struggles with poverty and a history of ethnic strife between its Kurdish residents and the national government. Van was the only city on our journey where we ever felt threatened. A young man on a crowded street yanked Janet's sunglasses from around her neck that evening and ran off.

The next morning, we rode the bus past the lake's rugged southern shoreline, crossed the scenic Taurus mountains, and descended through the upper Tigris River watershed to the city of Batman, a gritty oil production center. Our objective was a few miles farther south, the ancient village of Hasankeyf on the Tigris River.

It was sunny and warm the day of our visit to Hasankeyf, a town with dwellings, chapels, and a fortress carved into limestone bluffs along the Tigris River (37°42'N). Our guide that day was Ipek Tasli, the young and energetic local coordinator for the "Initiative to Keep Hasankeyf Alive," part of a coalition of 72 groups (municipalities, environmental organizations, and chambers of commerce). Hasankeyf is threatened by water that will back up behind Ilisu Dam, which is under construction a few miles inside Turkey's southern border. If completed, the 400-foot tall dam will create a massive reservoir and cover 200 villages, displacing up to 80,000 people. All but the highest sections of Hasankeyf would be submerged.

The setting, among rolling hills and sheer honey-colored cliffs rising from the edge of the river, is breathtakingly beautiful. This may be one of the oldest continuously inhabited townsites on Earth, with archeological evidence dating back 12,000 years. Hasankeyf served as a commercial center along the Silk Road trade route from China during the early Middle Ages: Two stone piers and an arch are all that remain of supports for a wooden bridge, built in 1116 A.D., that Marco Polo probably crossed.

We used a modern bridge, then hiked uphill to see some of the caves (there are thousands) cut into limestone cliffs for residences. Ruins of a hilltop castle built by Byzantine Emperor Constantine perched 600 feet above the river.

At Hasankeyf, on the Tigris River, one of the oldest towns on Earth, only the top of the minaret tower will remain above water if the reservoir fills behind the Ilisu Dam, now being constructed

Mongol's destroyed most of the castle. A secret stairway cut inside the cliff afforded access to river water during sieges. Hasankeyf eventually became part of Alexander the Great's kingdom and later was incorporated into the Roman Empire.

In a tea shop in one of the caves, overlooking the main street, Ipek told us her work for the Initiative began two years earlier, though the group formed in 1997. The Ilisu Dam is part of the Turkish government's Southeastern Anatolia Project, commonly referred to as GAP, from the Turkish language project name: *Güneydogu Anadolu Projesi*. GAP is one of the biggest hydroelectric dam projects in the world. Planning began in the 1950s, and construction in the 1980s on the $1.5 billion project. Twenty-two dams are being built along the Tigris and Euphrates Rivers, with 19 hydroelectric power plants.   The water is projected to double Turkey's irrigated farm acreage.

Our recently published maps were already out-of-date in depicting new reservoirs. Few free-flowing stretches remain on the rivers. As we traveled by bus farther west, passing one reservoir after another, each dam backed up slack water almost to the base of the next.

International resistance to the Ilisu Dam project developed because the reservoir will flood extremely significant historical and archaeological sites.   Swiss, German, and Austrian underwriters within the European Union had funded initial work, but pulled out in December 2008 after so many issues emerged. A list of 150 World Bank conditions dealing with the environment, protection of heritage sites, relocation of residents, and impacts on Iraq would have to be addressed if they were to provide funding.   The Turkish government, however, vows to fund the construction on its own, with completion projected for 2017. Ipek's coalition wonders if hidden money from the outside is making that possible. The local Kurdish population is wary, considering this dam additional pressure against the region's Kurdish communities and culture.

"Here is not Turkey," Ipek told us. "Of course, it is in the territory of Turkey, but we are not Turks. We are a different nation [that includes] northern Iraq, part of Iran, and part of Syria." She worries about possible arrest for opposing a state project, which is a crime in Turkey.

Turkey's dam-building and the reduced flows down-river from the Tigris and Euphrates Rivers are generating tension with neighboring Syria and Iraq. The two downstream nations demand that more water be released, but Turkey remains focused on filling its reservoirs. Forty percent less water will flow to Syria and even less to Iraq, farther downstream. Both nations depend on the Tigris and Euphrates river flows to grow food and serve growing populations. It is yet another stress on the political stability of the Middle East.

Despite these background concerns, the mood in Hasankeyf was celebratory on the day we visited, which was a national holiday. Returning to the riverside, we watched people of all ages splashing in the shallows and young adults dancing and singing in the street. Seated on divans around a low table in a restaurant built over the flowing water, we ate a meal of barbecued fish, fresh from the river, while looking down at other fish nibbling at our crumbs.

The Tigris River creates a unique riparian ecosystem. The local economy is based on fishing and farming. If the reservoir is filled, conditions for native fish will change and the river ecosystem will be gone. A local soft-shelled turtle (*Rafetus euphraticus*) has become a symbol of the impacts for a Turkish environmental group, *Doga Dernegi*, which maintains an information booth at Hasankeyf. The turtle used to be found along the Euphrates River, but is nearly gone there now because so many dams have been completed. Ilisu Dam may bring the species to the brink of extinction. Turkey has no equivalent law to the United States' Endangered Species Act, nor, apparently, a public trust doctrine on which to rely for legal protection. It could help if Turkey was accepted into the European Union, which has more progressive environmental laws.

That some people seem to care more deeply about these environmental concerns than about impacts on the local people, amazed Ipek. Though only 25 years old, she was very eloquent and knowledgeable. Her dark eyes blazed as she told us that a German Ambassador to Turkey has said that the nation is big, has lots of other cultural sites, and Hasankeyf's earthen cliffs and caves will be destroyed by time anyway. Despite such unfortunate "If you've seen one ancient site, you've seen them all" attitudes, she and her group remain determined to lobby internationally for pressure on the Turkish government. "Hasankeyf does not belong to Turkey, it belongs to humankind," Ipek said. "We ask the people of the world to save Hasankeyf."

We stood knee-deep in the Euphrates River after a long bus ride through the hills of eastern Turkey, westward from Hasankeyf to the city of Malatya. The Euphrates is the longest river in southwestern Asia, rising in highlands northwest of Van and terminating far to the south at the Persian Gulf. The fertile land between the lower Tigris and Euphrates Rivers cradled Mesopotamian cultures, and water from both rivers enabled the first irrigated agriculture, some of the world's oldest known human settlements, and the dreams and follies

Knee-deep in the Euphrates river

of kings. Only a few miles to the east, archaeologists were uncovering a 7,000-year old city called Arslantepe.

We waded into the warm water at a city park beside a ferry dock. On the way from Malatya, the local bus passed through Battalgazi, an ancient caravansera lodging place for Silk Road travelers and their camels. The Euphrates is one of the world's most famously historic rivers, though it was backed up there in yet another GAP reservoir, the Karakaya. Tamarisk trees lined the shore and provided shade in the park. They are an exotic problem species for desert wetlands in the United States, but part of the natural order there.

Suddenly we realized that we were surrounded by a half-circle of *gendarma,* the local military police, all carrying machine rifles. "What are you doing?" one asked. The English speaker was a handsome young man doing three months of national military service (we would learn), who taught English in Malatya and had been an exchange student in Seattle, Washington.

"Is it a problem to be here?" we asked. The answer was no, and though their demeanor was friendly-seeming, being confronted at the edge of the water by rifle-toting soldiers still made us uneasy. We exchanged a worried glance, wondering if going to places not frequented by tourists was a mistake that had finally caught up with us. Pulling out our map, we explained the book project that had us traveling around the world along that local latitude line.

The soldiers were there to meet a passenger ferry that arrived at the dock a few minutes later. One had a narcotics-detection dog on a leash that splashed into the water and lapped a drink. The ferry provided local service; internal security is a priority in this Kurdish section of Turkey where tensions have led to violence.

Once they finished inspecting the arriving passengers and vehicles, the young solder returned, saying that his commander invited us to come to their barracks in Battalgazi, to have tea and talk about their area's historic sites. We were not sure what to think of the invitation. Was declining really an option? We explained that we were using the public bus system.

They told us they could arrange a car ride for us. We walked with the translator to a sedan that had two men in civilian clothes in the front seat. They had been in the park when we arrived, perhaps undercover. Neither of those men spoke a word or even turned to look at us. Our car raced by the slower truck full of soldiers. Again, we looked at each other, silently wondering just what was coming next. The car passed through town and then through a gate into the *Battalgazi Gendarma Komutanligi* compound. After the gate swung shut, we were politely escorted to a nearby patio and served tea and mulberries. The commander, another officer, and our translator-soldier joined us (we asked everyone's name and also whether we could take their photographs, but were only allowed to record the name of Commander Samim Paksoy and that of the barracks).

They were very gracious people, quite interested in our book project. On a map, we showed them our travel route across Turkey, and they pointed to their home towns in central Turkey. They were proud of the nearby Arslantepe archaeological site and presented us with a book about mound excavations there, where an Italian group has been at work for 40 years.

Our worries proved a bit ridiculous, yet it was a big relief when our translator escorted us to a nearby bus station. When the public bus arrived, he spoke to the driver about dropping us near our hotel in Malatya, and when we arrived, the driver would not accept any money for the fare.

Malatya is famous for its apricots and the colorful bazaar where they are sold. Kamal, an ageless, effervescent character with long curly hair around a bald dome, holds court in his travel "office" – a table at an outdoor cafe in the park, and arranges tours to Nemrut Dagi.

On that 7,000 foot mountaintop south of Malatya, a 1st century B.C. king named Antiochus I (who ruled Commagene, a kingdom bounded by the Euphrates River) built a monument to Roman and Persian gods, placing his own statue among them. Fine sculptors shaped the likenesses of Apollo, Zeus, Hercules, Tyche and, of course, King Antiochus himself. The magnificent sculpted heads were forgotten by history until they were rediscovered in 1882 by an archaeologist. Major restoration was completed in the early 1980s.

Nemrut Dagi is one of Turkey's must-see treasures, though it takes considerable effort to reach the mountain. We opted for an overnight trip to a lodge that nestles a mile below the summit. The remote summit of Nemrut Dagi National Park lies almost exactly on our line, at 37°59'50"N.

We visited the mountaintop terraces and sculptures at sunset and sunrise. The massive figures once stood 25 to 30

First century B.C. ruler, Antiochus, had his own figure sculpted to place with the heads of gods on Mt. Nemrut Dagi. Small stones heaped atop the summit may cover a burial chamber

feet tall. Roughly hewn bodies still stand, but heads had fallen from above before they were unearthed. They still have a captivating, mystical quality. Honey-colored stone reflects the changing light, and the setting, on a conical summit crowned with small stones that may cover burial chambers, is magnificent. Shining through the morning mist, we could see another reservoir, far to the south, formed by Ataturk Dam on the Euphrates River, a reminder of the water source that irrigated the crops that generated the taxes that supported Antiochus's kingdom and made his extravagant monument possible.

UPDATE 2020
Ancient Hasankeyf began to be submerged in 2020 as the reservoir behind Ilisu Dam was 20 to 25 percent full that

February. Many residents have been relocated to the new town, Yeni Hasankeyf, on a hill across the river. A massive tomb, an ancient Turkish bath, the historic mosque and its minaret have been moved there also. Guards and dirt piles at both ends of the modern bridge (which will be submerged in time) block the way to Hasankeyf.

In 2015, war in Syria brought in Kurdish militias from southeastern Turkey and northern Iraq. Thousands of refugees fled Syrian violence. Though that was the primary threat driving the exodus, the crisis emerged following an extreme drought from 2006 to 2011, when water shortages prompted two million people to relocate to Syria's cities. *Time* magazine quoted Richard Seager, a climate scientist in Columbia University's Lamont-Doherty Earth Observatory, in a 2015 article about perceived connections between water shortages and the refugee crisis: "We're not saying the drought caused the war. We're saying that added to all the other stressors, it helped kick things over the threshold into open conflict. Global warming is desiccating the region in two ways: higher temperatures that increase evaporation in already parched soils, and weaker winds that bring less rain from the Mediterranean Sea during the wet season." (Baker, 2015).

Fairy Chimneys, Tuz Golu, Travertine Terraces, and a City That Lost its Port

A full-day's ride from Malatya on local buses, in the Cappadocia region of central Turkey, volcanic ash, hardened into rock and eroded by water through the ages has created a fantasy landscape. We left the high mountains and river canyons behind, dropping down to agricultural valleys before following a twisty road into the fairy chimneys, towers, and mushroom shapes that surround the village of Goreme. A huge, snow-capped volcano stood sentinel in the distance.

The whitish rock formations are locally called "tufa." In our many years of telling visitors about Mono Lake's tufa towers – the photogenic limestone formed when springwater and lakewater mix – we often heard about tufa formations in Cappadocia, though, technically, they are volcanic tuff. People carved cave houses and chapels into the formations and many were occupied, with gardens and apricot orchards nearby adding to the Hobbit-y feel of the place. Visitors may even choose "cave hotels" complete with windows, terraces and satellite dishes. Entire underground cities in multiple layers provided refuge during long ago raids and even horse stables had been carved into some of the larger rocks.

Three days there was barely a beginning for exploring the lovely area. Most of our time was spent hiking, seeking the high viewpoints to look down and across the valleys. Many tourists seek similar views in the dawn balloon rides that contribute significantly to the local economy. Wildflowers abounded, with Turkey mullein (another exotic pest at home that is native here) and irises decorating the

85

Fairy chimneys and columns of volcanic tuff provide amazing geological scenery in Cappadocia

hillside. A fox showed itself for a moment. Camels were there too, as a reminder of the historic Silk Road connections that Turkey has with China.

From Goreme, we made a day trip to find Tuz Golu, which means "salt lake." A huge white expanse shimmered in the heat, but it proved surprisingly hard to reach the shore of Turkey's second largest lake. From its southern end at 38°30'N it extends north for 50 miles. "Hypersaline" is barely descriptive enough; at 33 percent salinity, this is one of the saltiest lakes in the world.

The water that filled Tuz Golu that spring was fast evaporating with the warm temperatures. The lake appeared tantalizingly close, but proved a long slog through farm fields and wildflowers to the "shore" with the water itself still far across Tuz Golu muck. In winter, the lake's deepest point is only six feet deep. Every summer it goes almost completely dry, leaving behind a thick layer of salt that is commercially harvested and the source of 60% of the nation's salt supply.

Despite being so huge, the lake's high salinity limits life. There were no waterbirds to be found that day, though the largest Turkish breeding colony of Great Flamingos nests on a group of islands in the southern part of the lake. We could see the islands, doubled by their reflections on the water. Land birds did appear as we walked: storks and harriers flew overhead while many small birds flitted in the fields. A startled jackrabbit bounded out of the grass. We wondered when it had last seen a human visitor. Our approach to the lake was definitely not developed for visitors, though Tuz Golu's water and surrounding wetlands have been declared a specially protected   area.

Leaving Cappadocia's strange landscape, we boarded another public bus heading west on modern highways that follow the ancient Silk Road caravan route into western Turkey. Some caravanserai locations – those "truck stops" for camel caravans – still serve transport and travelers. The driver stopped for gas in Sultannan, where a caravansera was established in 1229 A.D., 3,000 feet above sea level. Above the high plain where we were traveling, Mount Hasan towered in the distance, a conical snow-capped volcano (10, 673 feet high), and one of the regional sources for volcanic deposits that make Cappadocia so magical.

Turkey has the largest fleet of bus companies in the world, needed in a large country where few people own cars and gasoline is expensive. In the east, buses were older than in the western tourist zones, ran less frequently, and were more crowded. Even the older buses were surprisingly comfortable. There was always a steward on board offering water, tea, and lemon hand freshener. The air conditioning usually worked. In the west, buses were fast and immaculate with nicer snacks.

Bus depots were clean, complete with young men or boys that greeted you with a cup of tea as you stepped off the bus. (Tea is ubiquitous in Turkey. The famed Turkish coffee is rare and expensive.) At bus stops, men smoked outside while the women disappeared into the restroom, where (Janet learned) headscarves came off and everyone lit a cigarette, away from the gaze of men.

One memorable ride began with a huge refrigerator being loaded under the bus. Later, the driver stopped at a roadside market where twenty cans of gasoline were laboriously loaded, one by one, into the gas tank, while passengers smoked cigarettes a few feet away.

Tufa towers are the signature visual element at Mono Lake, so it was interesting to learn that the 38th parallel in western Turkey encounters spectacular features made of calcium carbonate – the travertine terraces of Pamukkale National Park(37°55'N). Brilliant white terraces and aquamarine water draw tourists from around the world. The face of the travertine hillside looks out over orchards and wheat fields near the Buyuk Menderes River (which twists so much it gave us the word "meander"). Lukewarm spring water at Pamukkale is loaded with calcium carbonate which precipitates out as travertine when exposed to air, in a process similar to cave stalactite formation. It is another example of how calcium, carbonate and water combine to form beautiful features around the world.

The white terraces must be dry to really gleam, so the flow of warm spring water (95° Fahrenheit) is shifted by park staff throughout the day, which also provides changeable bathing experiences. The low-tech system includes ditches, sheets of metal and plastic bags. The challenge for

Pamukkale's travertine terraces and cliffs form where calcium carbonate precipitates from warm water in western Turkey

Pamukkale's managers is to protect the terraces while, at the same time, accommodating thousands of bikini-clad tourists that want to frolic in pools. People are allowed to swim in artificial pools constructed along an old road routed through the natural terraces. A crew armed with brooms was cleaning dirt and algae from the area that was off-limits.

Adding to the scene are spectacular Roman ruins above the travertine hillside, where Heirapolis was founded in the second century B.C. Multiple earthquakes eventually forced the Romans to abandon the city. A well-preserved ancient theater overlooks Roman baths where modern visitors still float above broken columns on the bottom of the pool.

Our day in Pamukkale was an interesting mix of wading, people-watching, and archeology, punctuated by the shrill whistles of guards whenever people ventured too far up the terraces or neglected to take their shoes off – bare feet are required for everyone, including the uniformed "rangers." The white hillside was lit up after dark, as frogs in the lake below croaked and warbled songs unlike any we had ever heard from a frog.

At the west coast, Turkey's grand archaeological jewel is Ephesus, which nestles in a canyon in a pine forest (37°56'N). During its heyday, between 100 to 200 A.D., 200,000 people lived in this Roman capital of Asia Minor. Notables from history such as Cleopatra, Alexander the Great, Saint Paul, and the Virgin Mary are tied to the history of this great city. Though founded as a harbor-town, with a man-made port at the mouth of the Cayster River, that critical tie with the outside world eventually ended, dooming the city because a flawed harbor design caused river sediments to accumulate. Today, from the top of the huge amphitheater, the sea is barely a gleam in the distance, five miles away. Hints of the city's watery history remain, with a "Harbor Boulevard" (paved with marble), harbor gymnasium and baths, and wetland marshes where the port used to be. At the edge of the marsh, on an old church site, several little frogs hopped across the stone floor.

The Celsus Library of Ephesus, built in 117 AD, is breathtaking, even when crowded with tourists off the cruise ships that dock, today, in Kusadasi. Terrace houses, replete with frescoes and mosaics are being pieced back together like giant jigsaw puzzles. Our new Turkish friend, proprietor of the Liman Hotel beside the modern port, said that what impressed him most about Ephesus was how they constructed buildings "without any sticky stuff" (mortar) or modern measuring instruments.

One of the seven wonders of the Ancient World was the nearby Temple of Artemis, just outside Selcuk, that served travelers stopping to pay homage to the goddess on their way to nearby Ephesus. An early version of the temple, constructed about 550 BC, was destroyed by fire, and a grander replacement begun in 356 BC. It was still under construction in 333 B.C. when Alexander the Great came to Ephesus.

Antipater of Sidon named the world's seven wondrous sites, about 140 BC, in a list that praised the Temple of Artemis above all the others:

> *I have seen the walls of rock-like Babylon that chariots can run upon, and the Zeus on the Alpheus, and the Hanging Gardens and the great statue of the Sun, and the huge labor of the steep pyramids, and the mighty Tomb of Mausolus; but when I looked at the house of Artemis, soaring to the clouds, those others were dimmed. Apart from Olympus, the sun never looked upon its like.*

Antipater, XCI, 69 (Gow and Page, 1968)

Only one 60-foot high column remains to give a hint of the 127 marble columns in that great Temple (the Parthenon of Athens has 86 columns). The single column remaining incorporates dissociated fragments discovered on the site in the 20th century. Storks were nesting on top when we were there.

Artemis was a fertility goddess for the people of Ephesus. Her statue in the nearby museum in Selcuk has multiple breasts (or maybe they represented eggs or even bull testicles, according to the museum staff). The site is marshy; Pliny the Elder wrote that such wet ground was preferable as protection against earthquakes. In 391 AD the Roman Emperor Theodosius the Great closed the Temple after declaring Christianity the new state religion.

Ephesus, on Turkey's west coast, was a Roman seaport until the river mouth silted in; today the coast is miles away, barely visible from the top row of seats in the ancient theater

The contrast between hordes of cruise ship tour groups at Ephesus and the lonely Temple of Artemis was striking. We wandered alone in the marsh at the end of an unmarked road that once held one of the wonders of the ancient world. Imposing monuments can, it turns out, have fleeting lives, making those that still remain along the marbled streets of Ephesus that much more precious.

We had crossed 750 miles of inland Turkey, by this time, and a dominant impression was how surprisingly green the country appeared. We had envisioned Middle Eastern aridity, in our ignorance. Instead, almost all of the land out the bus windows was irrigated and cultivated, except where highways ascended mountain ridges. Turkey is bountiful enough to feed its population and still have extra produce and grain for export, though the country's key rivers have been tamed at the expense of history and nature. Our westward progression through Cappadocia and Pamukkale had also shown us the role of water in spectacular natural landscape features and historical sites that are major attractions for travelers. Until Cappadocia, we had seen almost no other international tourists since leaving the Terracotta Warriors site in Xi'an, China, thousands of miles to the east.

UPDATE 2020
In 2017, a controversial project was announced to again connect Ephesus to the sea by recreating the ancient canal and allowing yachts to sail in and dock near the ancient port city.

Part II: Europe

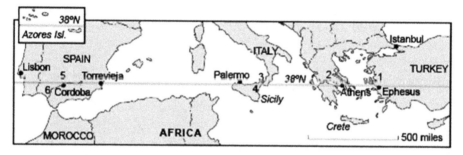

1. Izmir   2. Delphi, Greece   3. Messina   4. Mount Etna
5. Sierra Morena mountain range   6. Guadalquivir River

"Never think that a fight cannot be won. Never, never."
Anna Giordano

Greek Islands, Athens, and the "Navel of the Earth"

The west coast of Turkey borders the Aegean Sea and Greece, a nation with thousands of islands and a territorial span that is mostly water. Even on the mainland, one is never farther than 100 miles from the sea. Ferries are the key transportation mode connecting Greek islands. Most serve cars and trucks along with passengers and feel like small cruise ships, capable of smoothing out the Mediterranean's wind-tossed waters. High-speed catamarans, solely for passengers, are also available, with indoor seating that feels like travel inside a very wide airplane.

Approaching on a ferry through the Mediterranean basin's perennial haze, islands often appear to be capped by white snow that transforms into white-washed buildings perched on the clifftops. Though the Mediterranean Sea is often an intense turquoise-blue under the 38th parallel sun, the sky we saw was usually white. We wondered if the haze was simply seasonal humidity, but learned that particulate pollution travels south from Europe to settle in the basin. Throughout the Mediterranean Sea region we were also struck by a lack of sea and bird life, compared to the California coast. The sea is seriously over-fished. Much of the seafood common on restaurant menus must now be grown in aquaculture "farms." People have applied pressure to this environment for a very long time and it shows.

A Turkish ferry from Kusadasi took us in the back door to Greece via the island of Samos (37°46'N). Along the town waterfront, a door stood open to the Samos Ecological Society. Inside, we met three very friendly, helpful people, though only

Michaliadis Michael spoke English. The others kept smiling and nodding and answered our questions as Michaliadis translated. On the wall were posters about World Wetlands Day and also about Monk Seals (*Monachus monachus*).

The group works on education and environmental issues, particularly the rare, extremely endangered seal. Samos is one of the Monk Seal's last strongholds. Declining fisheries impact the seals' food too, and they are sometimes directly killed by illegal fishing with dynamite (one dead seal had washed up a few weeks earlier). There are also concerns about uranium in the seals' habitat, left by military missile exercises.

Exploring the west end of the small island, we hiked above a seal protection zone, scanning with binoculars for little heads in the sea. Alas, the rare Monk seal remained elusive.

On the south side of Samos island, near the village of Pythagorio, the Eupalines water tunnel was cut through a mountain over 2,500 years ago. The Mediterranean climate pattern, with winter rain and a long summer drought, creates water supply issues across Greece (and the entire Mediterranean region) that are very familiar to Californians. This 3,432-foot water tunnel was hand-chiseled through the mountain to bring water to a thirsty city of 80,000 people (today, the total island population is 32,000, with the biggest towns on the north coast). Working from both ends, the two crews met unerringly in the middle. In 524 BC they had no surveying equipment or compasses, but used a series of right triangles to calculate distances and stay on course, though Pythagoras (whose theorem about right triangles is familiar from high school geometry) was not born here until many years later. These were major mathematics and engineering feats.

What an intriguing connection for us: a thirsty city reaches out and diverts surface water into conduits that pass through a tunnel bored beneath a mountain, then to an aqueduct to reach the city -- a phenomenal engineering feat for its day and an example of just how much toil some leaders will set in motion (on Samos, by a despotic ruler, Polycrates, who used slave labor) to solve a perceived water supply problem. History repeated itself near our 38th parallel home, when the city of Los Angeles reached into the Mono Lake Basin to divert streams and passed that water along to the Los Angeles aqueduct through an audacious engineering feat, tunneling 11

The 2,500 year-old Eupalines water tunnel on Samos

miles beneath the Mono Craters mountain range, to avoid being constrained by water limits in the city's own natural watershed.

The ancient city of Athens is one of the largest cities in the world along the 38th parallel (the famous Acropolis in the city center is at 37°58'N). From Athens, our mainland travel route would circle the Gulf of Corinth, taking in ancient Delphi – the "navel of the Earth," the Peloponnese site of Olympia, home to ancient Olympic games, and the city of Corinth.

Athens was founded around the Acropolis hill, where steps to the Parthenon have been smoothed by thousands of feet for thousands of years. The Acropolis, like many Greek hills, is porous carbonate rock where water can dissolve its way along fissures and form caves, sinkholes, and natural "pipelines," often emerging as springs farther downhill. Springs and seeps from the limestone Acropolis were used by Mycenaeans as early as the thirteenth century BC. "The Acropolis is, in effect, a huge reservoir." (Fagan, 2011, 162)

The annual summer droughts of a Mediterranean climate brought water challenges to the ancient city as it grew (a seasonal problem faced across the region). Cisterns stored water and fed street fountains, some of which are still used. Once local wells could no longer handle the demand, aqueducts were built to transfer water from nearby mountains. During Roman rule, in 140 A.D., the Hadrian Aqueduct was completed from the foot of Mount Parnitha to Hadrian's Reservoir. As in Samos, routing aqueducts underground was intended to protect their precious supplies from enemy attacks. Hadrian's Aqueduct included a 15-mile long tunnel chiseled through rock, a tremendous feat for slaves using hand tools, with air shafts at approximately 100-foot intervals. Pipes distributed water through Athens, and drains handled sewage and storm runoff.

When Athens absorbed several million refugees from Turkey, in 1923 at the end of the Greco-Turkish War, the city's water needs became critical. About 26 miles to the north, along a route that today passes through planted fields, vineyards, olive groves and lush hillsides, is the city of Marathon. On the Marathon plain, in 490 BC, the Greeks defeated an invading Persian army. That battle location is remembered today for the long-distance run made by Pheidippides, a messenger who delivered news of victory to Athens and then died of exhaustion. Near Marathon, a reservoir was completed in 1929 that provided most of Athens' water supply into the 1950s. The same construction project included renovations to the ancient Hadrian aqueduct system, which still served the city. Marathon Dam, which backs up water from the Charadros and Varnavas rivers, is uniquely surfaced with white marble from the same source used on the Parthenon.

A more recent post-World War II population boom moved Athens to add Yliki reservoir water to its supply in 1959, using pumps to move water across hills northwest of Marathon. In 1981, the Athenians reached 120 miles farther west to the Mornos River, and built there one of the tallest earth gravity dams in Europe (375 feet high). The Mornos River's natural termination point is in the Gulf of Corinth.

The landscape and vegetation northwest of Athens and on the Peloponnese Peninsula looks much like California's coast range. Blue-domed churches in each village and signs in the Greek alphabet were the primary giveaways that this was Greece. The country is hilly almost everywhere, with many canyon creeks going dry in August. Away from the capital city, ancient agricultural lands support olive orchards, vineyards, and irrigated vegetable crops. Olives and grapes are well-adapted crops for the Mediterranean climate pattern, but hot, dry summer days present familiar water challenges for plants and people. Most roofs in Greece sport solar hot water panels, and dual-flush water conserving toilets were common where we lodged.

Water has played a special, mystical, role in Greek culture through the ages. The ancient Greeks associated natural fountains and springs with gods and goddesses, extolling their medical or mystical powers. The perfect example is Delphi, south of Mount Parnassus, where we learned that "the navel of the Earth" is a feature of the 38th parallel (38°29'N).

At Delphi's Temple of Apollo. a sacred spring called Castalia flows into the Cephissus river. Castalia was one of the naiad nymphs, goddesses who, in Greek mythology,

The Temple of Athena at Delphi, was home to Greek's prophetic Oracle. Delphi was considered the "navel of the Earth"

inhabited and protected rivers, streams, lakes, marshes, fountains and springs. The nymph Castalia aided the Delphi Oracle. Pilgrims posed questions for the Oracle priestess, who sat beside a cleft where sweet-smelling vapors rose from a spring below ground. She entered a trance, then recovered to deliver answers from Apollo, the god of prophecy, healing, and the sun.

Modern analysis of the spring in the sanctuary has detected ethylene. In low concentrations the sweet-smelling gas can induce a trance state, but exposure sometimes causes violent reactions, including wild, incoherent cries and thrashing – behaviors that fit historic descriptions of the Oracles' sometimes erratic behavior. Christian emperor Theodosius I, who shut down the Temple of Artemis in Greece, also ended operations at the "pagan" sanctuary at Delphi in 395 A.D.

Northwest from Delphi are the Messalongi marshes, the largest wetlands in Greece. We vowed to return some winter at the peak of the bird migration. The largest natural lake in Greece, Lake Trichonis, is to the north. Since California is struggling with problems of non-native zebra mussels introduced to lakes and streams, it was interesting to learn that this lake is part of the mussel's native habitat.

The bus trip between Delphi and Patras, the port city at the western mouth of the Gulf of Corinth, followed the shore so closely that our bus tires were in seawater when waves swept across the beach road. Patras is Greece's second largest city and a major port, roles played by San Francisco in the state of California. It even has a dramatic bridge spanning the "gate" into the gulf, called the Rio-Antirrio Bridge (38°08'N). Rows of cables on the recently completed bridge stretch from four towers, looking like harp strings. We had our cameras poised for the crossing, but just before we reached the bridge our bus turned down to an old ferry terminal. Puzzled, we asked why and learned that the toll is three times higher on the new bridge, so public buses still use the ferry.

Since leaving Athens, olive trees had covered nearly every hill, but south of Patras flat farmland grew corn and melons. Not far to the west, in the Ionian Sea, the island of Zakynthos has six beaches that serve as the most critical nesting areas for endangered loggerhead sea turtles (*Caretta caretta*) in the

Mediterranean (37°47'N). Volunteers with Archelon, the Sea Turtle Protection Society of Greece, work at those beaches during the nesting season, from early May through mid-August, educating about 70,000 tourists each year about the turtles' need for undisturbed beach sand. Loggerheads migrate long distances to reach the island, coming ashore at night to lay eggs. Key challenges are with tourist development and enforcement of laws that limit hours of beach activity and numbers of beach umbrellas and sunbeds. Since the turtles use the back edges of the beaches, human activities are limited to the beach sand closest to the water. The organization even convinced the government to stop night flights from the island airport, because the noise and lights confused turtles on a nesting beach in the flight path. They successfully lobbied for the establishment, in 1999, of the National Marine Park of Zakynthos and operate a sea turtle rescue center at Glifada, south of Athens, to treat injured turtles and serve as an environmental education center.

Turning away from the coast, our bus climbed back into hilly country on the Peloponnese Peninsula toward Olympia, site of the ancient – the original – Olympic Games (37°38'N). At that remote location, where the Kladeos and Alfeios Rivers meet, we finally saw a truly blue Greek sky. The ancient stadium sits quietly, enlivened only by tourists posing for photographs as if starting a race on the track, until torch-lighting ceremonies return there every four years to launch each modern Olympic Game. Greece was not united when the original Olympics were happening (776 BC to 393 AD). When each athletic competition was held, a truce was declared between participating warring city-states. A 38-foot tall statue of Zeus made of gold and ivory, erected at Olympia, became yet another of the seven wonders of the ancient world along the 38th parallel (though it disappeared toward the end of the 4th century AD). Dedicating the Olympic games to Zeus did not sit well with Christian Emperor Theodosius I, of course, so he ordered the events to end.

Back on the bus, heading east toward Corinth, our ears popped as we wound up and down rolling hills, following an ancient highway route through the Alfeios River canyon and a landscape reminiscent of the Sierra Nevada foothills. In Greek mythology, the Alfeios was one of two rivers Hercules re-routed

101

to cleanse the Augean stables. The road narrowed where it passed through old towns and whenever oncoming traffic met our bus, horn blowing and complex maneuvering ensued to negotiate which vehicles would have to back up and pull aside. The length and windiness of this highway helps explain why this part of Greece remains undeveloped and little disturbed by tourism. In a place where an unoccupied car blocked the road, our driver's horn brought a woman from a nearby house. She came to his window, spoke awhile, then walked back into the house to find her keys, cheerily waving toward the people on the bus. There was no
grumbling while we waited; a lady so exceedingly pregnant is accorded great respect in Greece.

The Corinth Canal cuts through the narrow Isthmus that separates the Peloponnese Peninsula from the Greek mainland, providing a shipping shortcut from the Gulf of Corinth to the Aegean Sea. In 67 AD, Roman Emperor Nero put several thousand workers to digging, but the canal was not completed before his death. Work stopped and did not resume until 1881. Finished two years later, the 26-foot deep, four-mile long canal is only 79 feet wide, too narrow for many modern ships, yet still a busy conduit for commerce.

The fortress of Acrocorinth is high above local farms and greenhouses and the Gulf of Corinth

We were met at the Corinth bus station by Nico, who keeps an apartment for tourists beside his house and runs a tavern in town. He drove us up the hill to the Acrocorinth fortress, the oldest in the Peloponnese (37°54'N). This had become a notorious site by the time Paul's first letter to the Corinthians in the Bible warned the citizens against visiting harlots there. Fortress walls show modifications by a succession of outside rulers: French, Byzantine, Venetian and Turks. At the summit, there was an expansive view of the Gulf of Corinth that we had circled during the prior week. Greenhouses dominated the ground between the hill and the sea; Corinth is especially famed for the currants grown there.

The Acrocorinth hill is, like the Acropolis, a porous limestone feature with many springs emerging along its northern base. Legend has it that a woman named Peirene was transformed into one of those springs by tears she shed after her son was killed by Artemis (bringing to mind that goddesses' temple in Turkey). The ancient gods and goddesses spent considerable time, apparently, along our line of travel.

On our last night in the country, back in Athens, we checked into a hotel after dark and walked to the Parthenon. The ancient temple was lit up, glowing high above this extraordinary city on the 38th parallel.

In Italy, the 38th parallel passes through the south end of the mainland, then closely follows the north coast of Sicily from Messina to Palermo. A ferry ride from Calabria brought us to Sicily across the Strait of Messina.

UPDATE 2020

Since 2015, refugees, most fleeing war in Syria, have flooded onto Greek islands, including Samos, where drinking water supplies already required deliveries by tanker ship. Greece established refugee camps on the mainland, but the numbers became overwhelming and the society has struggled to handle the influx.

The climb to the ancient Acrocorinth fortress

Saving Migratory Raptors in Sicily

Negotiations between water and land created the narrow Strait of Messina, which funnels raptors migrating up from Africa, annually seeking the shortest route across the water. From Italy, the birds spread out toward nesting sites in northern and eastern Europe. We volunteered at an annual raptor-counting camp sponsored by the Mediterranean Association for Nature and the World Wildlife Fund. About 45 volunteers help out each year, from mid-April to mid-May, the peak migratory season.

The annual bird count has a dramatic history, with protective surveillance against poaching a key part of the effort. Anna Giordano is the spark plug who has fired up international volunteers here since 1984. That year, as only 3,198 raptors flew by, the team heard 1,185 gunshots. Birds were maneuvering through a deadly gauntlet. Thousands of migrating raptors were shot by local Sicilians every year, though the birds had been legally protected since 1977. The senseless massacre was a traditional hunt for trophies; stuffed birds were displayed in homes and presented as gifts to show off shooting skills. A local folk legend even held that any man who could not kill at least one Honey Buzzard (hawks are called "buzzards" in Europe) during the spring migration would be cuckolded by his wife.

We made it to the camp, a Forest Service facility, as it got dark (38°16'N). It was windy and cold that night. Volunteer counters come and go each day during the counting season; ten were there when we arrived. Jean Paul Fiott from Malta, had worked on the count for ten years. Michele Cento from Rome, works with the Italian League for Protection of Birds and was counting for his fourth year. It was the first birding adventure for Sophie Tyldesley, an 18-year old English girl. In the prior three days, the counting team had tallied 12,000 raptors, smashing the one-day count record for April, so spirits were high.

International volunteers county migrating raptors at the Strait of Messina in Sicily and report illegal poaching

Our day was spent on a windy 1,200-foot ridge in the Peloritani Mountains, with a view of snowy Mount Etna spewing ash in the distance. Groups of raptors kept arriving from the southwest, circling on thermals to gain altitude before attempting the water crossing. The ridge had been a particularly hot spot for poaching in the past. As Anna Giordano told us her incredible story, her eyes constantly scanned for approaching birds. She usually picked out raptors before anyone else and quickly identified the distant profiles. Each time she would stop talking, confirm that the sighting was recorded by the team, and then take up exactly where her narrative had been interrupted:

"Since I was six years old, I had been a member of the Italian League for Protection of Birds. I was always in love with nature. When I was 15, we had a country house near Messina and I heard about the migration of raptors in the Strait and that they were being shot by hundreds of poachers. I had known birds migrated at Gibraltar, and the Bosporus Strait, but to think they were also by Messina, where I lived, no."

It seemed incredible to her that the birds were being shot, because hunting was prohibited in the spring time. All raptors had been legally protected since 1977. She decided to see if what she had heard was true.

"Near here, I met a poacher that wanted to give me, as a present, three kestrels he shot. Of course, I refused. I saw many birds killed, falling down into the bushes. I found shotgun shells with 'honey buzzards' printed on them in the local dialect. And then, one day, I was surrounded by about 35 poachers. They became very aggressive..."

She paused to look at approaching birds.

David interviewing Anna Giordano on a ridge above Messina. She founded the annual count to help end the killing of migrating hawks

"...and everyone had pistols, guns, knives. They invited me to go with them to a 'meeting,' and said, "What would your parents think if you do not come back home? Go! Do something else."

"They were threatening you?" we asked.

"Yes," she confirmed, then had to answer a phone call from her office about more volunteers arriving that day. Hanging up, she said: "That was the first of their last days. I was very, very angry. I couldn't tolerate that they were shooting protected raptors from illegal hides [blinds] and nobody was doing anything. I was upset and angry at the same time and started this long campaign that has taken 29 years."

"You were only a teenager and yet your parents never discouraged you?" we asked.

"They never expected me to *not* do something, because they knew, if they said 'no,' I would do that thing." She laughed, recalling herself as a teenager. "They have the psychological approach to push me the right direction. Those first years, 1981, '82, and '83, I could do almost nothing out here. Then I started receiving great help from the Malta Ornithological Society and from Germany, and began this annual camp."

Another arriving bird flock interrupted her.

"In the beginning we stayed as close to the shooting as possible, in that bunker (she pointed to a concrete shooting hide, down the hill), but after [the BBC released the film, *Anna and the Honey Buzzards*] some poachers stopped letting us use the bunker. From 1985 to 1990, we used my mother's house for the camp, then we moved because she was tired of having 40 persons in her house." Anna laughed, but her next words clarified the nature of the battle she had taken on: "No one would rent us a house. They were all afraid that some poacher could burn or do some other damage to their houses."

They finally found a house to rent on the ridge, then later in a village we could see down in the valley, until that became too expensive. Being able to use the Forestry Regional camp for the last two years signaled a change in government agency attitudes that they had long sought.

"At first the government did not support us. They were thinking: 'Let them shoot. There are many other problems in the world.'"

There was another pause for the approach of birds, this time Herring Gulls.

"So, in the beginning there was not too much help. When there was poaching, I would race all the way down the mountain to report to the Forestry or *Carabinieri* [police], driving like crazy. Sometimes I would come into the office and the man was reading the newspaper, drinking his coffee, and..."

Again, a pause for birds.

"...he just told me that they don't have time to come. So I said, 'If you don't come immediately I will write a charge that you are not doing your job.' So, they began coming, but in the beginning, there were no arrests, because they didn't have the will. But they were coming.

The first arrest for poaching finally happened in 1984. "That day I was completely crazy, because they were shooting everywhere. I went for the *Carabinieri*. We started..."

She stopped and pointed. "Look there, a honey buzzard. Well, we started down the hill. The poacher saw us and started to run. So, then, I made a mistake; I shouted and ran after him. I found him hidden in the grass, but holding no gun and no birds, and when I turned my head, I saw that I was alone with this man. The Carabinieri had not followed. But I was not hurt, perhaps because he had hidden his gun, perhaps because he was an official, a traffic warden. Finally, the Carabinieri arrived, and a friend that was with me found the gun and dead honey buzzards."

"A scary moment!" we said.

"Yes! The man was arrested and began telling me, 'Oh, please, I am the father of a family. Please don't destroy me.' And I was thinking, yes, he is father of a family. Then I looked at the honey buzzards' black bills and said "These were also the head of a family.'"

Anna was a young female challenging a male hunting tradition in a Mafia dominated society. Her car was firebombed in 1986; fortunately, she was not hurt. Her house was broken into. A dead falcon was left at one of the observation points with a warning note: "Your courage will cost you dearly." Only after shots were fired at her and some of the volunteers did the police change their do-nothing approach.

And, gradually, local attitudes improved and her efforts began to pay off. Anna first felt things were changing for the better in 1989, when she read in the newspaper that, without her prodding, Carabinieri had arrested a poacher for illegally shooting quail. The first year the counting team heard no gunshots was 2005. Shots are now a rarity, though toward the end of our day we heard two down in the valley as a flock of hawks circled above the distant farms. Anna and Jean Paul made a cell phone call to the police and raced off, but were unable to locate the shooter. When she returned, clearly angry, Anna declared, "Apparently 29 years is *still* not enough."

In 2006, something occurred that, Anna said, "gave me a sense of why I have devoted my life to birds. There was a man, one of the worst poachers and worst men I ever met in my life. He was always very aggressive. One day he stopped here for the first time and he hugged me and said, 'Can we talk?'

They moved a little away from the counting group and the man said, "I remember what you told us 25 years ago."

And Anna was thinking, she said, "My god, what did I say 25 years ago? I didn't remember."

"You told us," the man said, "that one day we would start loving these birds and not shoot them anymore. I just came to tell you that you were right."

"I could not even find words," Anna said. "It gave me goose bumps. So, I gave him some binoculars. Before, when he was going out from the farm, he would ask his son, 'Do you have the gun?' And now when they go out, he asks, 'Do you have the binoculars?'"

In one respect, Anna's story reminded us of the 16-year battle to save Mono Lake, during which a few key people proved to be so important (though they never faced such a violent response).

"I am convinced that each one can make the difference," she said. "At the same time, alone you do nothing. I had only the responsibility to start the song that became a chorus."

"But somebody had to sing that first note," Janet pointed out.

Anna could not take such direct praise without making a self-deprecating joke: "Ah, I have a genetic deviation in my mind that makes me seek the justice that here does not exist. Next time, I will be born with a passion for the stamps." She laughed. "My life is dedicated to a fight and any more is too much. I am joking, but in my next life, maybe I could be a judge so I could affect the people with power."

We asked Anna what she would most like us to write, what she most wanted to tell people.

"Never think that a fight cannot be won. Never, never."

More birds passed overhead to be identified and counted, and then she added: "Our role now is to prevent poaching. Before, it was to repress poaching; now to prevent. This Strait of Messina is a treasure of biodiversity. We have a high amount of birds, fishes, insects, plants, sea mammals, and many endemic and rare species. If I became rich one day, I would like to make an international observatory; that has always been my dream. And a Secret Service for the environment," she added, laughing. "But I'm not able to find the funds. This is my fault. I can find a poacher but I'm not able to go to a rich man and say, 'Oh, trust in us.'"

A BBC television documentary, *"Anna and the Honey Buzzards"* appeared in 1992, and she received the prestigious Goldman Environmental Prize in 1998, which annually honors grassroots environmental heroes from each continent. *Time Magazine* told Anna's story in 2000. We told her we would also write about her effort and maybe people will help her desire for an international observatory come true.

By the time the 2010 count finished, 35,397 birds were tallied, including 31,124 Honey Buzzards, 1,769 Marsh Harriers, 583 Common Kestrels, 520 Black Kites – a total of 28 raptor and stork species.

Anna's courageous campaigning extends beyond the raptor count. She operates a wildlife rehabilitation center in Messina and was fighting to stop construction of apartments and condominiums on unstable mountainsides when we met. In October 2009, 37 people had died in mudslides. The Mafia seemed to be laundering money by building unsafe housing. Because of her public opposition, she was having to fight eviction from the building that housed her rehabilitation center. She is also combating plans for a bridge over the Straits of Messina that, she feels, could have devastating environmental impacts. The middle-aged dynamo's work is not yet over.

We moved westward across Sicily on a very slow train, stopping at each local station, then caught a bus from Palermo, the island's largest city, to Trapani and from there flew to Spain.

UPDATE 2020

Raptors continue to migrate over the Messina Strait and to be counted by volunteers. We have lost touch with Anna, but efforts to create a Messina Strait bird observatory continue. In 2014, the European Court of Justice acted against the Italian government for not adequately treating water sources contaminated with calcium, arsenic, and fluoride. Most Italians choose to drink bottled water instead of the supply from their taps. In 2017, Italy saw its most severe drought in 60 years and a severe heatwave, a pattern consistent with climate change forecasts.

Spain's Coastal Lagoons, Water for Growth, and Iberian Lynx

The 38th parallel encounters the Spanish coast at Torrevieja (38°00'N), a town famed for its salt works and singing competitions. There, at 00°42'W, we were also west of the Greenwich Meridian, zero longitude line for the first time. Two immense saline lagoons cradle the town of Torrevieja, one colored green and one pink. At La Mata, the green lagoon, we saw Brine Flies and bright red Brine Shrimp (*Artemia salina*), Eared Grebes (*Podiceps nigricollis*), osprey and avocets – species that are common at Mono Lake, but these lagoons were below sea level, the ocean was close by, and there were no snowy mountain peaks in the background.

The pink lagoon, colored by iron-fixing bacteria, is hypersaline. It tasted intensely salty, like the Great Salt Lake in Utah. Park ranger Antonio Saez explained that the two lagoons were natural features managed for centuries for salt production. A canal lets in the sea when the water level gets low. Machinery scoops salt off the lagoon bottom into huge drying piles. A colony of the rare Audouin Gull (*Gaviota audouin*) uses the salt works as a nesting site.

Both lagoons attract wintering and nesting waterbirds, most notably over 3,000 Eared Grebes, Northern Shovelers (*Anas clypeata),* Red-crested Pochards (*Netta rufina).* and up to 2,000 Greater Flamingos (*Phoenicopterus ruber).* Viewing blinds were spaced along a trail at the edge of the green lagoon, where a helpful English birder shared his spotting scope. A Norwegian choir was singing in the picnic area (providing an unusual background sound for birding, but not an uncommon sight around Torrevieja).

The largest lagoon on this coast is a few miles south of Torrevieja. The "small sea" of *Mar Menor* provides more than 36,000 acres of bird habitat for nesting, migrating, and winter refuge (37°49'N). A long boardwalk helps people reach several swimming basins. Stretching out on our backs and floating buoyantly in the knee-deep salt water reminded us of swims in Mono Lake, but mud bathing was added to this experience. It was strangely enjoyable to spread smooth black mud over our bodies, scooped from the bottom of the shallow lagoon.

This part of the Spanish Mediterranean coastline is called *"Costa Blanca."* Most of the white, sandy beaches are backed by mile after mile of thirsty condominium resorts and golf courses. Southern Spain boasts over 300 days of sunshine a year and draws sun-seekers from northern Europe to *Costa Blanca, Costa Brava, Costa del Sol,* and even *Costa del Golf.* Outside the resorts, imported water irrigates flowers, fruits, vegetables, olives, and vineyards that flourish in the warm weather.

Spain's water issues are very similar to California's. There is a geographical mismatch between available water and demand. Long-distance aqueducts extend from the wetter north to serve urban and agricultural thirsts in the arid south. In 1973, the Tajo River was diverted southeast from central Spain across 175 miles to the Segura basin in Murcia province. Our drive to *Mar Menor* passed farm fields outside the city of Murcia and crossed a bridge over the Tajo-Segura aqueduct. The project was primarily to serve farm irrigation and solve groundwater depletion problems, just as California's Central Valley Project was meant to stop farms from overdrafting the southern San Joaquin Valley aquifers. But in both cases, irrigated acreage expanded using the new water while aquifer problems persist. Fertilizer and pesticide use also increased in the Segura River and it gained the dubious distinction of being the most polluted Spanish river. In Spain as in California, water has also been shifted away from farms to serve the seemingly unending growth of tourism and urban development.

In 2004, when a new government took power in Spain, one of its first acts was to scrap another water transfer plan. Ebro

River water was going to be taken from northeastern Spain, where there are 19,700 acres of natural wetlands and another 52,000 acres growing rice. The controversial project threatened the Ebro Delta environment, which had Special Protection Area status for birds, was part of the Natura 2000 Network in the European Union, and also a Natural Park.

The new government's alternative, *Programa AGUA*, (*Actuaciones para la Gestión y la Utilización del Agua*), did not rely on long-distance water transfers. The focus became sea water desalination: 24 plants to generate 50 percent of the water in the AGUA Program. Another 20 percent is to come from wastewater re-cycling, 15 percent from modernizing irrigation systems, and 15 percent from other efficiency measures.

The largest desalination plant in Europe was under construction at Torrevieja during our visit, near the east side of the pink lagoon, but work seemed stalled. Our requests to the construction company for a tour of the plant were never answered. The guard at the gate of the complex just shrugged and rolled his eyes when we asked when they might finish. The ranger's comment back at Mata y Torrevieja park was: "*Muy* [very] *political.*" A woman staffing the tourism visitor center told us that the project was controlled by the national government, so local people did not know much about it.

Developing sufficient renewable energy sources (the preference in the AGUA plan), to meet energy costs (including the costs of desalination and additional costs to pump water inland, uphill), and mitigating the environmental damage to the ocean when concentrated brine is deposited, are challenges that plague ocean desalination efforts everywhere. Europe's financial crisis was a complicating factor. One desalination plant was already operating near the *Mar Menor*, at San Pedro.

While planning the Spanish portion of our trip, we thought of Javier Grijalbo, an old friend from Janet's exchange-student days, 35 years earlier, in Madrid. We reconnected with him by letter – a real one with a stamp and everything – explaining our project. Javi is a renowned illustrator and naturalist, the perfect guide. He and his wife, Virginia, traveled with us for five days through Andalucia, a place they know well, identifying every plant and bird (and all the good restaurants).

The Sierra Morena mountain range closely follows the 38th parallel across most of Spain between Andalucia and Andújar (38°07'N). Water drains down its southern slopes into the Guadalquivir River, which flows past the historic cities of Córdoba and Seville, then into a huge estuary, *Las Marismas,* before entering the Atlantic Ocean.

The dehesa of Spain is semi-wild woodland of olive and cork oaks where livestock roam the watershed of the Guadalquivir River

Much of the mountain landscape is *dehesa* – semi-wild woodlands where olive trees and cork oaks are cultivated, but also managed for livestock grazing and hunting grounds. Black-footed Pigs (*Pata negra*) root for acorns beneath the oak trees and provide the finest ham in Spain. The eastern half of the Sierra Morena range also provides wildland refuge for the last stand of the Iberian Lynx (*Lynx pardina*), perhaps the most endangered wild cat in the world. They are relatives of our North American bobcats, but smaller, standing about two feet tall, with tufted ears and sideburn whiskers framing their faces. Only 176 lived in the wild, when we visited, with another 67 in a captive breeding program on the southwest coast inside Cota Doñana National Park. With Javi and Virginia, we would go to both locations to learn about Spain's efforts to save the lynx.

North of Andújar, in the Jándula river basin, the four of us spent a day with German Garrote Alonso, who works for the environmental and conservation arm of the regional government. When he spoke, Janet could immediately hear that he was from Madrid, the accent she best understands since her student days in that city.

The day before, a half-million people had swarmed into this region on an annual overnight pilgrimage to a hilltop shrine. German took us through locked gates into a 5,000-acre project area that is off limits to the public, into the wildest landscape we would see in Spain. There he works with wild Iberian Lynx and tries to expand European Rabbit (*Oryctolagus cuniculus*) numbers, because they are the critical food source for the lynx.

Rabbit disease epidemics are the major cause behind the decline in lynx. Myxomatosis, a viral disease of South American origin, was deliberately introduced into Australia in 1950 and to France in 1952. It spread into Spain, quickly killing off 90 percent of the rabbit populations. As had happened in Australia, after an initial plummeting of numbers, the Spanish rabbit population showed signs of recovery as individuals with natural resistance to the virus increased. Then, in the 1980s, rabbit hemorrhagic disease arrived. The total reduction in rabbit numbers exceeded 80 percent during the last 30 years. Road kills and hunting pressure added to the problem. Rabbit hunting is one of the most traditional hunting activities on the peninsula.

It has not proven easy to increase the rabbit populations everywhere, so an alternate tactic brings lynx to remote lands preserved for deer hunting, like the Sierra de Andújar, where rabbits still thrive. Of three pairs of lynx relocated the winter of 2009-10, one female had given birth to three cubs and another was pregnant when we visited. But some determined cats (their locations monitored with GPS) frustrated researchers by heading back to their original territories, traveling hundreds of miles.

German can individually recognize all of the lynx older than a year by their appearance, after years studying photographs taken with motion-sensor cameras. He puts out supplementary feed for the lynx once a week. He showed some of the cameras, strapped to trees, and several of his artificial rabbit warrens, built inside enclosures as an experiment to increase rabbit numbers.

Since the program began in 2003, numbers of lynx in the wild have increased by 40 percent, but it is still too early to be certain of success. As German said, "We have in our hands the power to restore the lynx. They are flexible and will respond. They just need to have the chance with protection of their habitat."

Outside the enclosed area, after we said goodbye to German, the four of us quietly waited and watched in the late afternoon near the Jándula River, a place where sightings are possible. We saw neither lynx nor rabbits. *"No importa, están allí,"* Javi later wrote in his blog: "No matter, they are there."

The following January, our naturalist friend went back to Sierra Morena and saw lynx four times. Through a spotting scope, Javi watched a lynx stalking, then pouncing and digging, rapidly throwing dirt, pausing to look around, clawing at the ground again, until it pulled a small rabbit from its burrow which it voraciously devoured.

Spanish people support the campaign to save the Iberian Lynx, though very few have seen them. Warning signs along the roadsides near Andújar prominently displayed a lynx face and the single word: "Recuerde" – Remember. At a store in the lynx region we noticed a decal with a painted lynx image and showed it to Javi. He did a double-take. The artwork had been pirated from one of his beautifully crafted educational posters. Javi vowed to correct that misuse of the image on commercial products without his permission.

The Jándula River is one of many tributaries draining from the Sierra Morena toward the Guadalquivir, which becomes Spain's fourth longest river and one of Spain's largest as it passes the ancient city of Córdoba (37°53'N). Between 200 B.C. and 400 A.D., Rome controlled the city and made it the capital of the Roman Province in southern Spain. Roman ships navigated from the Atlantic Ocean all the way upriver to Córdoba (boats today travel up only as far as Seville). They built three aqueducts to deliver water from mountain streams into the growing city.

A Roman bridge crosses the Guadalquivir River toward the Mezquita mosque in Córdoba

Arab Muslims ousted the Romans in the 8th century A.D., and constructed the *Mezquita*, Córdoba's great mosque, on top of a Christian basilica. Under the Caliphate of Córdoba from 929 to 1031, the city became the most populous in the world (200 years before Merv topped that list), and was considered the intellectual, political, and economic center of Europe – accomplishments that were made possible by ingenious solutions to the water supply problem. Along the Guadalquivir valley, during the al-Andalus period of Islamic dominance, thousands of waterwheels lifted river water into an extensive system of irrigation canals, similar to those used for the same purpose in Lanzhou, China.

Walking past gates on the city streets offered glimpses of inner patios, gardens, and fountains, and the lush use of water that typifies life in the city. The weather was sunny and hot and it was a relief to return to the banks of the river, where new walkways had been completed since our last visit in 2003. Back then, we had noted a painted message on the wall above the riverbank: "*Rios Vivos, No Canelización*" ("Living Rivers, No Channelization"), a sign of the watershed's environmental challenges and local population concerns.

The route Javi and Virginia laid out for us passed through part of La Mancha province and across Andalucia. Washington Irving, in his book, *Tales of the Alhambra,* compared the maturing of a girl to womanhood as, "passing from the flat, bleak, uninteresting plain of La Mancha, to the voluptuous valleys and swelling hills of Andalusia." (Irving, 1994, 146). We looped back into the Sierra Morena range north of Seville, which is designated as "natural park" land. Development is limited there and the *dehesa* woodlands are protected while the traditional agricultural economy and village life are also preserved. Pigs and cattle graze beneath cork oak and olive trees. We saw a Short-toed Eagle (*Circaetus gallicus*) with striking yellow eyes, a single Black Stork (*Ciconia nigra*), and many White storks (*Ciconia ciconia*).

Following small tributaries back to the Guadalquivir River, then downstream past the city of Seville, led us to Coto Doñana National Park. Marshes at the estuary where the river spills into the ocean are renowned for water birds. The great river deposits sand and nutrients along the Atlantic Coast, creating extensive dunes, and scrub and *Pinus pinea* (pinyon pine) habitat on higher land. The region seemed unusually lush following an exceptionally wet winter in Spain, and those wetter conditions actually made it harder to see birds, since they could spread out across more of the verdant landscape. Javi and Virginia's great birding knowledge guided us to each habitat type, where they pointed out Flamingos, Spoonbills, Egrets, colorful Bee-eaters, and animated English Chaffinches (*Fringilla coelebs*), very common there, but new and fascinating to us.

Human activities around the edges of the park include charcoal production, bee keeping, wood gathering, fishing, and livestock grazing. Coastal condominiums and resort facilities lined beaches, similar to what we had seen on the east coast of Spain. Much of the scrub habitat on private land was converted to intensive irrigated agriculture (especially strawberries), causing groundwater levels to drop and interfering with the natural cycle of seasonal floods across the marshes.

There were also threats from the upper watershed. Pollution came down the Guadalquivir River when a wall at a toxic waste reservoir collapsed at a mine at Aznalcollar (37°31'N) in April 1998. Millions of cubic yards of toxic mud and acidic water washed down tributaries of the Guadalquivir, contaminating farmland along the riverbanks and continuing to the main river channel, a plume that, fortunately, stopped just outside the national park.

Doñana is the western habitat refuge for Iberian Lynx. We visited the Iberian Lynx Captive Breeding Center in Acebuche, inside the park, and met with Director Antonio Rivas. While the "in situ" project we had explored in the Sierra de Andújar focuses on relocating wild lynx onto appropriate habitat, the "ex situ" breeding program at Doñana aims to build up captive numbers for future reintroduction. Nine cubs were born there in 2010. The breeding adults and cubs are absolutely off-limits to visitors. We were not allowed to let them see us or to see them, except on video screens, to ensure they remained leery of people. It was a privilege to see the action on video screens in a monitoring room staffed night and day. The ambitious effort is comparable to the successful captive breeding program for endangered California condors. The plan calls for a core of 60 to 70 thriving individuals before introducing adult lynx into three habitat zones: Andújar, the western Spanish province of Extremadura, and eastern Portugal. That number had actually been reached when we visited the center, though none of the captive-born cats had yet been released. Two females bred in captivity with a wild-born male, finally did make it out, in February 2011, to prowl for rabbits on the wildlands of Andújar.

During our days together in Spain, Javier's list of bird sightings grew past 120 species. We parted after securing their promise to visit Mono Lake soon, and traveled into the province of Extremadura, the "extremely hard" land of western

Spain that produced so many of those extremely "hard" conquistadors who explored the New World. The city of Jerez de los Caballeros (38°19'N) was the birthplace of Hernan de Soto and Vasco Nunez de Balboa. From there, we crossed the border into Portugal.

UPDATE 2020
In 2016, after thousands of dead fish washed up on the shore of *Mar Menor*, 55,000 people marched in protest in Cartagena. Due to runoff from farms, golf courses, industry and housing, the lake turned green with extreme eutrophication, and most of its bottom life died, including a population of endangered fan mussels. Land management policies were blamed.

In 2019, the European Union reproached Spain for not controlling illegal wells around the Doñana National Park. The Water Alliance of the Guadalquivir River took action, shutting down 120 wells used by fruit farmers.

Flamingos migrate from Africa to the estuary of the Guadalquivir River and other wetlands in southwestern Spain

Portugal's Transported Town, a Solar Donkey, and the Azores

Outside the eastern Portugal town of Moura (38°08'N), where we appeared to be the only tourists, we visited the Alqueva *barragem* ("reservoir,") on the Guadiana River. The rolling hills were shaded by olive trees and carpeted with wildflowers of every color. The dam itself is not particularly huge, but the reservoir behind it is the largest in Europe. The new reservoir, completed in 2002, filled for the first time during the winter of 2009, a wet season that followed years of drought. No water deliveries had been made from the reservoir, yet, because distribution canals were still being built. Like California's Central Valley Project and the Tajo-Segura system in Spain, the Alqueva water will primarily go to farms.

The small village of Luz was in the path of the rising water, so its 300 residents were forced to move to a whole new town built on higher land. Several displaced residents told us that the town provided by the government just was not the same, though the general layout was similar and their neighbors had not changed. They watched, bitterly, as the old town and its surrounding forests were razed before the reservoir filled. A movie in the new museum brought tears to our eyes with stories of old fruit trees, favorite rose bushes and gardens lost to the reservoir. Luz was a reality check on the human upheaval that often happens when waterways are dammed and residents uprooted.

"The day they closed the gates [on the dam], we brought a wooden "white elephant" to the ceremony," José Martins said, the symbol of an unwanted gift that is difficult to dispose of and comes with upkeep costs far beyond its usefulness or value. José is the regional director for Quercus (Latin for "oak"), which is the major NGO environmental group of Portugal. He spoke quietly as he educated us, in his office in the city of Beja (38°01'N), about many issues surrounding the Alqueva reservoir.

Portugal and Spain had an agreement dating to 1968 to divide the Guadiana River water, but Spain pushed ahead with more than 40 dams, large and small, mostly for hydropower. while Portugal waited until 1995 to start building its dam. The project was designed to store enough irrigation water for 250,000 acres, "but how are you going to have enough water for that, considering climate change?" José asked. "Here in the western Mediterranean there may be a lot less rain in the coming years. Also, we don't believe Spain will send us enough water in dry years to fill the reservoir, because Spain is using water on thousands and thousands of hectares of land. They have one dam almost as big as Alqueva."

Pipelines for water from Alqueva Reservoir will expand irrigated agriculture across eastern and central Portugal

Water quality, particularly the gradual increase of chloride salts in the soil, was another concern. Plant roots separate irrigation water from the dissolved salts it carries, leaving the salts behind, until soils become worthless for agriculture. Lost farming soils in the "Fertile Crescent" between the Tigris and Euphrates Rivers of the Middle East are an historic example; the salt problems on California's southern San Joaquin Valley farms are another.

"Another problem," José continued, "is that water is being transferred from one river basin into another. Most of the irrigated area is not in the Guadiana watershed, so less water is going down the river." That alters chemical and biological factors and no one knows, yet, what the effects will be on fish, algae, "or even bacteria; sometimes the little thing is the worst," he said. Water diversions will also reduce flows that transport nutrients and sand to the river delta, where it enters the sea, diminishing the life in that productive estuary. "But all this is taboo to speak about," according to José. "They say, 'Spain is doing it, building dams and transferring water between basins.'" The Portuguese government seemed determined to follow the example set by its larger neighbor.

In fact, thirst for irrigation water was partly driven Spanish corporations recently purchasing olive-growing lands inside Portugal. "They have the money. Money!" José exclaimed. "It's just business. They don't want to know about the environment, about rivers, about riparian forests. They grow olives that now look like small vines. Intensive; super intensive. They kill everything else. That is an environmental disaster, in the last 5 or six years, on more than 120,000 acres here."

Hardy, drought resistant olive trees evolved in the Mediterranean climate and were cultivated successfully by small-scale farmers for thousands of years. The complex ecological relationships and habitats of olive tree woodlands (known in Portugal as "*montado*") were being lost to the new intensive olive culture that forces higher yields with pesticides, herbicides, and irrigation.

The price for water in Portugal is set low for political reasons, despite a European Union directive that water be priced at its actual cost. We explained how, in California, farmers get water at subsidized prices to help keep the price of food low for everyone, yet they can sometimes market that water to cities at a big profit. José was flabbergasted.

"That's completely...I don't know what to say. That's crazy! Well, here we have not gone that far." But they have some "crazy" ideas of their own: "Some people believe, because we have the dam, it is going to rain a lot more here. Back in 1993 or '94, I remember, the President said, with Alqueva and more evaporation, maybe we can have more rain. I said, if that was right, would Spain need so many dams? Would places by the sea, like California, be so dry?"

The Quercus organization is only 25 years old. Portugal's long dictatorship had to end, José told us, before such an organization could exist. "Our mentality..." he paused, then said, "we don't have a tradition to join organizations, because of that political era."

Near one of the world's largest solar electric arrays, a donkey pulled a loaded cart, including farmer and son, near Moura, Portugal

José works on many issues, including nature protection, environmental education, energy, and climate change. Quercus only has a few thousand members and 40 employees. Most financial support comes from corporate donations. They host a national television show every day and a radio show that airs internationally, called, "One Minute for the Earth."

Quercus also owns and manages land for environmental education and a micro-reserve network to protect cork woods, orchids, and other species.

Toward the end of our conversation we discussed the world's largest solar electric generating array that we had seen near Moura. There were 92 individual panels in dozens of clusters. This was a central utility approach to power generation, though we agreed with José that it ultimately makes better sense for a nation to invest in solar panels on every house or business property. Outside the solar facility, an incongruous sight had appeared that day against the modern backdrop: a donkey cart driven by father and son with a load of hay.

"We have a good latitude for sun, yes?" José said, concluding our talk on an appropriate note.

On the Atlantic coast at Portugal's *Costa Azul*, we stayed in the small fishing town of Porto Covo. The birthplace of Vasco de Gama, the seafarer who rounded Cape Horn to established the ocean route to India, was nearby in the port city of Sines (37°57'N). Driving north up the coast, our destination was *Lagoa de Sancha*, a lagoon entering the sea that attracts up to 13,000 flamingos and many storks. Signs were posted about endangered Least Terns, another similarity to our California coastline.

Returning to Sines, we located the fort beside Vasco de Gama's historic family house and his statue, staring westward into the Atlantic. Nine hundred miles in that direction were the Portuguese Azores islands, our next destination.

The first thing we had to grasp in the Azores was the fact that São Miguel Island (the most populated island of the Azores archipelago) was *way* out in the Atlantic, 940 miles from the Portuguese mainland, yet still over 2,000 miles from the east coast of the United States. All islanders must be self-sufficient, but the Azores is particularly distant from everywhere (37°46'N; 25°19'W).

The existence of the Azores became known by the end of the 14th century. Medieval legends of Atlantis, a sunken land in the Atlantic, and of *Sete Cidades,* a fabulous Kingdom of Seven Cities, became associated with the islands. Diogo de Silves officially "discovered" Santa Maria and São Miguel

Islands about 1427. In the prior decade, Portugal's Prince Henry "the Navigator," had been operating a navigation school at Sagres, Portugal, to promote the study of the sea and geography, and to develop improved ships for long distance exploration.

During Columbus's first voyage, in 1492, his flagship *Santa Maria* grounded on a reef on the outbound voyage and sank. The smaller boats, *Niña* and *Pinta* began their return trip to Spain in January, 1493. A day after the two became separated during a storm, on February 15, Columbus reached the Azores island of Santa Maria (though it would make a nice story if the island name was tied to Columbus's first flagship, it had been called that ever since a lookout on de Silves' ship exclaimed "*Santa Maria*" when the island appeared on the horizon). The *Pinta* missed the Azores entirely and only made landfall after reaching the Spanish mainland.

On São Miguel, we were never more than a few miles from a glimpse of the ocean, a constant reminder that this island was isolated in a vast sea. That close relationship is reflected in the name given to a North Atlantic feature that shapes the climate of Europe's Mediterranean region: the Azores High. Just as California's Mediterranean-style climate is controlled by a North Pacific High that blocks Arctic storms during the summer, then slides south to open the "door" to winter storms, the Azores High controls the weather patterns of the western Mediterranean and North Africa. Each summer, the sub-tropical high (or anticyclone) keeps the mainland region dry and warm, and when the High shifts south in the winter, wet northern systems may again deliver rain to the Iberian Peninsula.

Another constant, on the Azores, was evidence of volcanism. Lava rock is commonly used for building. Hot springs and fumaroles abound. In places, holes dug into the steaming ground served as cooking ovens. We visited steamy, bubbling Furnas, the thermal park of São Miguel, where a huge pool welcomed swimmers. After time in the warm water, Janet's once white shirt will forever be a nice shade of orange because of the minerals. Geothermal power is a key energy source on these isolated islands. We stopped at the geothermal plant that provides 40 percent of the island's

*On the volcanic Azores, islanders cook meals in holes
dug in the steaming ground*

electricity, but as we walked through the open gate a man ran up to shoo us out. He became much friendlier when we told him that the plant looked a lot like the geothermal facility outside Mammoth Lakes, not far from our Mono Lake home. He knew about that operation, yet photographs were strictly forbidden as the design of their plant was proprietary information.

Sete Cidades only has one small village (not seven as the name implies) in the center of a huge caldera holding two lakes, one blue and one green. The local legend tells of a princess and a shepherd, lovers who were not allowed to marry. The tears they cried in different colors gave each lake its distinctive hue.

According to a fisherman at Mosteiros, on the island's farthest west shore, fishing boats only venture out when the sea is calm – and one never knows when that may be. There were natural bathing pools protected from the waves by lava ridges along a rocky shore and used by local swimmers when the weather is warm, though not during our chilly mid-April visit. In fact, a woman hiker told us that she had been warmer back home in Finland!

Some features of the island archipelago reminded us of the Hawaiian Islands. The Azores had lots of flowers, particularly azaleas. There were wet and dry sides to the islands determined by the prevailing winds, as in Hawaii, but this 38°N archipelago is definitely *not* at a tropical latitude (though pineapples grow there in greenhouses).

During our stay, we heard that Icelandic volcanic ash was disrupting air travel throughout much of Europe. We wondered how long we might have to remain on those lonely islands in the middle of the Pacific? It brought home the real isolation of the Azores. Luckily, our flight route was far enough south to avoid the ash and we departed on schedule for the United States.

UPDATE 2020

On March 6, 2020, two Iberian Lynx were released in the Guadiana Valley, bringing the number of the species living wild in Portugal to 109.

Water from the Alqueva reservoir and relatively cheap land prices are luring foreign farm investors, including almond and berry producers from California, to purchase farmland irrigated by the reservoir.

Part III: United States

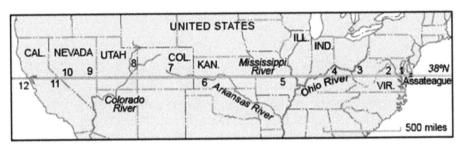

1. Chesapeake Bay   2. Charlottesville   3. Charleston, West Virginia
4. Lexington, Kentucky   5. Rolla, Missouri   6. Dodge City   7. Pueblo
8. Moab   9. Pioche   10. Tonopah   11. Mono Lake   12. San Francisco

"Economists are winning; not hydrologists."   Mark Rude
　　　　　　　　　on the overdrafting of the Ogallala Aquifer

Chesapeake Bay Watershed Education

Our east-to-west crossing of the United States began at the Atlantic Ocean and Assateague Island. We drove down the eastside peninsula separating Chesapeake Bay from the Atlantic Ocean, known here as "Delmarva," a poetic unification of the three states on the peninsula: Delaware, Maryland, and Virginia. At the Virginia state border, a sign read: "Dixieland, the South Starts Here." We passed through the local village on Chincoteague Island and crossed a bridge to wild Assateague Island and the Chincoteague National Wildlife Refuge. A lovely, shell-blessed beach stretches there for miles, home to endangered Piping Plovers (*Charadrius melodus*) and horseshoe crabs.

At the Refuge headquarters, biologist Joelle Buffa explained refuge management and, with volunteer historian Myrna Cherrix, we climbed to the top of the Assateague lighthouse – seven flights on a circular staircase – for views of the barrier island. Reaching the lighthouse required a walk through a forest where the mosquitoes were intense. Fortunately, Joelle provided mosquito jackets, with mesh and hoods to keep the mosquitoes from penetrating the fabric. We were casual, at first, about covering our faces with the netting, but learned fast.

The lighthouse was first lit on October 1, 1867. It towers above a forest now, because sand deposits built up new land around it. The south end of the island keeps growing, though the entire low-lying island may be covered by the rising sea in the next century, according to climate change projections. Meanwhile, the light still burns brightly at night. It is intriguing that the United States is bracketed by lighthouses

(Assateague and Point Reyes) and large bays (Chesapeake and San Francisco) with ocean outlets on the 38th parallel.

Within the wildlife refuge, there were many familiar birds: herons, Great Egrets, Canada Geese (*Branta canadensis*) and Mallards, but a Tri-colored Egret (*Egretta tricolor*) and Laughing Gulls (*Larus atricilla*) were new for us. Monarch butterflies migrate along both coasts. Sitka deer *(Odocoileus hemionus sitkensis)* wander the island; they are related to mule deer, but introduced from Japan. The Chincoteague pony is the most famous wild animal in this area.

The ponies are gathered, each July, to swim between the two islands, so that yearlings can be sold, a way to control the island population which has no predators, while raising money for the Chincoteague Volunteer Fire Department. The children's novel, *Misty of Chincoteague* by Marguerite Henry, was inspired by a true-life story of children determined to buy a pony, Misty, at the annual auction. Local legend claims ponies swam onto the island after a Spanish shipwreck. It is more likely that settlers first put them there to graze where no fences were needed to contain them.

The north half of the island is a National Seashore, operated by the National Park Service, and the south half is part of the National Wildlife Refuge. In the National Seashore, the pony herd is not culled by annual drives and auctions, but population size is managed by injecting mares each year with contraceptives.

Joelle invited us to spend the night at her home nearby where our GPS confirmed that she and her husband Clyde live *exactly* on our latitude line. Clyde retired after many years directing the Don Edwards National Wildlife Refuge in San Francisco Bay, and Joelle had previously been in charge of the Farallones Islands, so their move east had essentially followed the 38th parallel. We discovered a number of mutual west coast birder and environmentalist friends. Clyde had been a Yosemite ranger in 1976 when researchers came to study Mono Lake and got to know the founders of the Mono Lake Committee.

Chesapeake Bay is 195 miles long and up to 30 miles wide; it makes San Francisco Bay look small by comparison. The watershed has 50 large rivers and thousands of smaller streams and extends north to Cooperstown, New York, and inland into West Virginia. "Drains to the Bay" signs cautioning against putting pollutants down storm drains are found far inland.

We joined two school classes on environmental education boat trips that started from different points on the Bay. The first trip, from the east shore, required a ferry to Smith Island (38°00'N), 10 miles out into the bay, where the Chesapeake Bay Foundation operates an Education Center. For much of the trip we could not see land from the boat. The Bay felt like a huge calm sea.

Environmental education coordinator, Krispen Parke, hustled us directly over to another boat, one of the Foundation's fleet, loaded with students from Clarke County High School in Berryville, Virginia, west of Washington D.C. They were all from rural farms and "knew all about plows, cows, and sows," their teacher said, but this was their first experience on the Bay. Their goal was to better understand how agricultural practices in the upper watershed affect nutrient and sediment deposition and, ultimately, the whole Bay ecosystem. The program is the reverse of a strategy that brings school kids from Los Angeles to the Mono Lake basin to help them understand how their actions "downstream" in the distant city affect their upper watershed. One of the Foundation's guiding philosophies is that "the place to teach people about the Bay is on it and in it."

Off we went, back onto the Bay to drag the bottom and learn about its living creatures. Chesapeake's average depth is only 22 feet, but during that cruise we were usually in less than five feet of water. Even so, the murky water made the bottom invisible. The biggest problem in the Bay is nutrient deposition from agriculture and urban runoff, which fertilizes algae, causing bacteria populations to increase when the algae dies and decomposes. That consumes oxygen and creates "dead zones." Sedimentation also smothers the grass.

Overfishing has devastated the marine ecosystem. Fishermen used to harvest enormous numbers of oysters and crabs. Oysters clean the water as they filter it for their plankton food, and then fertilize the bottom with their wastes. "The oyster is a good example of how we sometimes underestimate the importance of creatures by considering them mainly as they relate to our bellies and our commerce," local author Tom Horton wrote (1991, 27). Yet, Foundation educators know that lessons often are best remembered when related to other senses, so the students also pulled up 17 crab pots, put out the day before, and collected dozens of soft-shelled blue crabs that would become their dinner that night. They measured each crab's shell and learned to tell males and females apart. Interestingly, there are no size limits on females, since their size does not relate to age. Oysters were put back into the water because the season had not yet started. An assortment of other bottom dwellers also showed up, including pipefish, grass shrimp, jellyfish, and grass that supports much of the bottom life.

The students had placed their crabpots along a state boundary line. Krispen told the story of Ben Marsh, an oysterman and storekeeper, who successfully lobbied to have the Virginia/Maryland border placed right through the Smith Island archipelago, so that whichever state fishing regulations were most favorable would apply.

On the ferry returning to the mainland, we met Tom Horton whose books, about the Chesapeake Bay Foundation's efforts and about raising his family on Smith Island, we had discovered during research for the trip. Tom has not lived on Smith Island since the 1990s, but was there that day with another school group. Protecting Chesapeake Bay is an overwhelming task, but, Tom said, "It would be a lot worse by now if we hadn't done anything."

He was referring to the Chesapeake Bay Foundation, which was established 45 years ago. "Save the Bay" bumper stickers are familiar across the watershed (a communication medium that also served the campaign to "Save Mono Lake" across California). The organization now has over 200,000 members and 160 staff members and can claim considerable success, yet in 2010, their annual "State of the Bay index," which monitors 13 key ecosystem indicators, stood at only 31 (a perfect score would be 100).

Our second day with a school group on Chesapeake Bay was on Captain Jack Russell's skipjack boat, another Foundation vessel that works out of St. George Island (38°09'N), where the Potomac River enters the west side of the Bay. We joined third graders from Beach School in Chesapeake Beach,

135

Maryland on a cruise that was an interesting contrast to the day spent with high school students.

Skipjack boats are rakish sailing vessels that were used to work oyster beds before Maryland laws allowed power-driven dredge boats. The activities on Jack's boat were familiar, though they *sounded* different on this side of the Bay. We "drudged" for "arsters," according to St. George Islanders, that are called "ersters over on Smith Island and the eastern shore; and "awsters" up in New England. Jack had shucked oysters for 30 years, but there had been no commercial product taken since 2000. He has shared his knowledge with more than 60,000 school kids.

Captain Jack Russell showed a schoolgirl how to hold a crab without getting pinched; her class was cruising to learn about Chesapeake Bay

Cap'n Jack's approach was perfectly adapted to third graders and it was delightful to watch a professional at work (since we had conducted tours to many school groups at Mono Lake). Before boarding, he had all the kids stand in a ring holding a rope, called a "line" by sailors. When he told them,

"You may get fish guts on you today," the kids cheered loudly (we wondered how that line would have worked with the high school crowd). Jack said, "Kids, y'all have to do *all* the work." That produced silence. "Parents don't get to do any." The kids yelled, "YES!" falling for the old Tom Sawyer whitewash-the-fence trick. As life preservers were passed out, each inscribed with the name of some part of the boat, the teacher had to raise her voice, "Girls, it doesn't matter what color they are."

Before launching we looked at a live eel and a Bardog in tubs. Jack explained that Bardogs, also called Toadfish (*Opsanus tau),* have no commercial value, yet were in decline, an indication of problems in the Bay ecosystem that go beyond fishing pressure. One mother, along as a chaperone, missed Jack's point, asking, "Isn't that good?" (apparently because the Bardog is commonly called a "trash fish"). Jack's simple answer: "Everything has its place."

When he asked the students what they thought they might see that day, one answered, "Gold fish?" "Well, OK," Jack said. "Some people do throw theirs away into the bay." We appreciated how he never discouraged participation by reacting negatively to "wrong" answers.

Ospreys nested on every channel marker post as we motored out. Jack kept talking, teaching us names for the parts of the boat, making sure to mention each student's life jacket name. He taught everyone how to tie bowline knots.

As nets dragged along the bottom, Jack explained the sorry history of overfishing and ecosystem decline that was now familiar to us. Oysters each filter 50 gallons of water a day and used to clean the bay's water, but numbers had slid downward ever since the 1970s. "There are too many people. Their wastes have to go somewhere, and that's right here," Jack said.

A few seconds after bringing the dredge off the bottom, Jack pried open an oyster and turned to David to show the kids how to eat it. They watched closely as it slid on down.

The Cap'n told us that oyster season was in the months with an "r:" September through April. Then he recited part of Oscar Nash's poem, "Oysters:"

I'd like to be
An oyster, say,
In August, June,
July, or May.
                    (Nash, 1931, 30)

We pulled up crab pots and Jack taught the kids how to hold a crab without getting pinched by its claws while throwing it back. Blue Crab numbers were the only bright spot on the Chesapeake, Jack said, "because crabs like trash and polluted water." While baiting the empty crab pots, Jack had the kids gather around for a solemn ceremony thanking the alewife bait fish.

After snacks were handed out, one boy asked the teacher, "Do we get recess now?" She told him, "This day IS your recess."

The overall message Jack wants to send away with the students is: "Place a piece of the Bay in your hands. Maybe you will be impressed enough with what you see to take better care of the it than my generation. It's our Bay...let's pass it on!"

One of the student letters posted on the wall in Jack's office reads: "I learned the left side of the boat is called port and the right side is called starburt [sic]. My favorite part was getting my hair wet. I loved the trip. Jenna." She drew a picture of the boat on the water, with the sun overhead wearing sunglasses.

UPDATE 2020

In 2018, for the first time in 33 years, Chesapeake Bay's health showed improvement in every region. Still, the Bay's overall health grade was a "C." The federal cleanup plan administered by the Environmental Protection Agency was credited for the improvements, but President Trump tried, unsuccessfully, to eliminate funding in 2018, and did cut it by 90 percent for fiscal 2019. Congressional support is the key in the six states involved, to push back against the Trump administration's anti-science, anti-regulation policies.

The Rappahanock River and Mattawoman Creek

To reach the west side of Chesapeake Bay, we had traveled south across the mouth of the Bay through a long underwater tunnel. Joelle, the biologist at Chincoteague, had suggested we visit a hawk surveying and songbird banding site along that route, near the end of the peninsula. Just as the Strait of Messina in Sicily funnels migratory birds, Delmarva concentrates migrating raptors, songbirds, shorebirds, some waterfowl, and even butterflies on their approaches to the Bay mouth. The Coastal Virginia Wildlife Observatory staff and volunteers have counted birds there since 1963.

On a calm, sunny morning at Kiptopeke State Park, near Cape Charles, we walked from our campsite to the hawk-watch platform. An average of 30 hawks, eagles, and falcons pass overhead every hour during the migration. Coopers and Sharp-skinned Hawks were hurtling on the wind straight south, rather than soaring in circles as in Sicily, seeming in a rush to make the crossing. There were lots of Yellow-shafted Flickers, too; the group had counted over 400 already that morning.

At a bird banding station in the nearby woods, Ann Gilmore and several volunteers gathered songbirds caught in mist nets to measure and band them. Janet got to release a Northern Pirula back to the wild with a tiny new band on its leg and wished it well on its long trip to South America. It was hard to believe such tiny bits of fluff, feather and feistiness could fly so far. Another treat was a cardinal in the hand, very common here, but strikingly red to us westerners.

After we made our own crossing, we stopped briefly at George Washington's Birthplace National Monument, where a bald eagle flew by along the bank of the Potomac (38°12'N). That seemed an appropriate and auspicious sighting.

In Fredericksburg, Virginia, we walked through an idyllic woodland with John Tippett, the slim, articulate director of the Friends of the Rappahanock River, to see the former site of Embrey Dam (38°19'N). In 2004, Navy SEALS blew the dam up, allowing the river to flow freely after many years of effort by John's group. The dam had filled with sediment and lost its usefulness for hydroelectric generation and as a water supply for Fredericksburg, because alternative ways to capture water upstream had been developed. John brought along a photograph taken where we stood, of the free-flowing river overlaid by a "ghost image" at the instant the dam exploded, telling us how exciting it was that day to set the river free. "The fireworks of a spectacular explosion is not the norm in environmental protection," John said, "which usually consists of lobbying, action alerts and fundraisers."

A nearby plaque commemorated the achievement: "Since the mid 19th century, the Rappahanock river had been holding its breath behind a wall of iron, concrete and wood. The rapids formed where the Embrey dam once stood are now and forever to be known as John W. Warner rapids," to honor the United States Senator who was a leader in the campaign. That positive story brought home to us, again, how often a key person at a certain moment in time (in this case, Senator Warner) makes a difference to a special place.

Until the dam was removed, anadromous fish, including herring, American Shad (a threatened species), Alewife, and Striped Bass could not complete their life cycles by returning from the bay to spawning grounds upriver. Through the 20th century, fish returning each spawning season swarmed in frustrated masses at the base of the dam. As part of the Embrey Dam campaign, each autumn, on the third Saturday in September, the Game and Fisheries Department had electronically stunned the fish long enough for volunteers to scoop them into buckets. People formed a human chain to move the fish and release them above the dam. After those fish spawned, their small offspring could wash over the dam spillway to return to the bay.

That effort, mostly symbolic, kept people involved in the years before their dam-removal goals were achieved. It reminded us of the annual Mono Lake bucket walks that delivered water from above the creek diversion dam down to the lake and helped inspire a generation of activists. The Friends of the Rappahanock still hold a festival on the third Saturday each September.

The water supply alternative to the big dam involves a totally different approach. Today, many small dams provide storage in side drainages that are filled by pumping water from the main river channel: Strict restrictions ensure that the pumps "only sip from the river when it's running at high flows," John said, to avoid impacting the river ecosystem.

The Friends group also helped secure city ownership of river-front lands on the entire Rappahanock watershed and 30 miles of permanent conservation easements. "That took three years, two city councils, partnerships with the Nature Conservancy and others," John said. An endowment also funds a "river cop" to enforce regulations along the river.

As in many places around the vast Chesapeake Bay watershed, the group's long-term battle is with sprawling development. Wherever paving and construction make land impervious to water, nutrient and sediment run-off increases. Chesapeake Bay Foundation scientists have documented a threshold – 10 percent coverage of the land – beyond which drastic declines in biological health occur in the Bay. "We've made progress on all of the major point sources, like wastewater discharges and agricultural waste, but have been going backwards when it comes to urban runoff," John told us. "In fact, it's counteracting everything else." Though a billion dollars has been spent to upgrade community wastewater treatment plants in Virginia, what runs off yards and parking lots is equally important, because of population growth and unchecked development.

We crossed the Potomac, leaving Virginia to reach Maryland's Mattawoman Creek, one of the most pristine rivers draining to Chesapeake Bay. The creek was number four on the American Rivers Organization's annual listing of America's most threatened rivers in 2009 (the Sacramento-San Joaquin River system, in California, led that year's list). It was no surprise,

141

by then, to learn that the biggest threats to the Mattawoman watershed were sprawl and development. Things were at a crisis point because a major cross-county highway proposed in the upper watershed would harden many miles of land and serve even more suburban sprawl.

Kayaking was the best way for Jim Long (left) to explain the threats to pristine Mattawoman Creek, from development on its Maryland watershed

Smallwood State Park campground is beside Mattawoman Creek (38°33'N). The "creek" name had us expecting a much smaller river. In Mattawoman's wide channel, fresh water is pushed back up the creek by each rising tide that enters from the bay. The campground was almost empty. We fell asleep listening to a Great Blue Heron's croaking call and geese honking overhead.

The morning dawned overcast but calm and we met Jim Long, president of the Mattawoman Watershed Society, and Bonnie Bick, vice president of the Society and the Sierra Club representative on the Mattawoman issue. The four of us kayaked up Mattawoman Creek, gliding among huge round pads of American Lotus, with Great Egrets winging overhead. The Mattawoman is the most pristine of the Chesapeake Bay tributaries, in spite of being just 25 miles south of downtown Washington D.C.

Proximity to the national capital is the root of its problems. The watershed of the seven-mile long river contains the fastest growing region in Maryland and had already reached the tipping point where more than 10 percent of land was impervious to water. The paving, concrete, and construction that would follow the proposed new highway could ultimately change water runoff on 25 percent of the creek's watershed. Nearly two-thirds of that new road was finished, leaving the stretch closest to Mattawoman. Permits are required to disturb wetlands, and Jim's organization was hopeful that national and state agencies would deny them.

We asked Jim, a bespectacled government physicist, what moved him to work so hard to protect the Mattawoman. "I started kayaking all around the Bay and eventually found Mattawoman," he said. "I realized how rare it is and it became my biological center."

On our return to the launch point, the national board of directors for American Rivers had a large group ready to begin kayaking. Jim and Bonny were to attend their dinner that night, hoping for national support for their local efforts.

Protecting Chesapeake Bay is a daunting challenge because runoff problems originate across such an enormous watershed. "We need to learn how to behave, and quickly," author Tom Horton wrote in 1991. "We have already ensured that nearly a generation will grow up in parts of the watershed without the ability to catch shad and rockfish; ensured that a generation or two will never wade as children in grassy, clear shallows in pursuit of softshell crabs; and risked that many generations to come will take congested highways and strip development and sprawling suburbia as their normal environment." (Horton, 1991, 288).

In 2010, the Environmental Protection Agency created new Total Maximum Daily Load objectives for the Rappahanock and the mainstem of Chesapeake Bay, aiming for a 25 percent cut in nitrogen and phosphorus and at least a 16 percent reduction in sediment. The states around the Bay must develop plans with specific two-year actions aimed at pollution reductions within the next 15 years. Also, President Barack Obama issued an executive order in 2010

143

that declared Chesapeake Bay a "National Treasure" and called for a strategy to restore and protect the Bay, its watershed and resources.

In November, 2011, two years after our visit, the Mattawoman Watershed Society celebrated news that the Maryland Department of the Environment had denied the cross-county connector highway project's wetland permit for failing to properly address its environmental impacts.

During one day of travel, the 38th parallel took us by President James Madison's Montpelier home, James Monroe's birthplace on the Potomac River in Fredericksburg, Thomas Jefferson's famous Monticello home in Charlottesville and, in Staunton, Virginia, to the birthplace of Woodrow Wilson. Jefferson was taken with the idea of growing grapes and olives at Monticello after visiting Italy, thinking that they might thrive there because of the shared latitude (38°01'N). He had limited success. A guide told us that Jefferson, who kept exact weather reports and was an inventor and amateur scientist, "was most likely very aware of his location on the 38th parallel latitude; it would be very Jeffersonian."

Atop the Blue Ridge Parkway, we stopped to look back to the east on an exceptionally clear day. The next high point along the latitude, back that direction, was a peak on the Azores islands. We left crabs, creeks, coasts, and curlews behind and headed west, where the continental United States awaited. Scarlet fall color added a touch of flame that grew more intense as we crossed the West Virginia state line (38°07'N).

UPDATE 2020

In 2017, Mattawoman Creek's future health received a reprieve when Charles County adopted a "Watershed Conservation District" plan to reduce development density on much of the creek's watershed.

West Virginia and Mountaintop Removal Coal Mining

Pocahontas County, West Virginia, straddles one of the continent's water divides, and is known as "The Birthplace of Eight Rivers," sending water back down toward Chesapeake Bay and westward toward the Ohio River, part of the vast Mississippi River watershed that drains the middle of the continent. At a festival in Hillsboro, West Virginia, at the historic birth home of author Pearl S. Buck, several volunteers enthusiastically named the eight rivers for us: the Greenbrier, Elk, Williams, Cherry, Cranberry, North Fork of the Cheat, Tygart, and the Gauley.

That same day found us at the Cranberry Glades in Monangahela National Forest (38°12'N). The Glades are strangely fascinating because Arctic bog species are found there, delivered long ago by an ice age glacier. Cold, snowy conditions at 3,366 feet above sea level, plus local acidic soils allow the plant community to persist far outside of the Arctic. Hues of brown and gold, with pitcher plants, cranberry moss, spruce, hemlock, and alder trees, enticed us to walk twice around the loop trail. If Thomas Jefferson had visited here he would have better understood that latitude is not the sole factor dictating plant growth, because micro-climates, topography, and soil types are so important.

A raft trip on the lower Gauley River was an exciting 14-mile whitewater experience near Fayetteville, West Virginia (38°04'N). Whitewater rafting has become an important alternative to the coal industry that dominates the regional economy. Some of the most popular rafting stretches in the eastern United States are along New River Gorge National

Rafting the Gauley River

River (one of only four "national rivers" in the National Park System), plus 40 miles of tributaries in the Gauley River National Recreation Area, where ten local guide companies serve about 200,000 people each year.

Our raft guide was Justin Reynolds, son of a coal miner, who had run the New River 200 times by the time he was 16 years old. A dam on the Gauley provides steady water flows for barges heading downriver toward Charleston, except when pulses are released to serve recreational rafting. Trips on the upper and lower Gauley are timed to ride the "bubbles" of water released from the dam on six weekends each fall. Our arrival during the short rafting season was serendipitous.

It was a wild ride with Justin, who sought out "surf holes" where the boat sat in one spot while the river crashed around us. When our raft sprung a leak, Justin had his hands full trying to steer as our anxiety inflated while the boat gradually deflated. Luckily, the river was busy with other rafters who loaned us a pump.

The owner of our raft company, Skip Heater, grew up locally and shared his knowledge about regional history over breakfast. A key reason for establishing the New River Gorge National River in 1978 was to stop a coal strip-mine planned for the gorge. In 1873, railroad tracks were extended along the river to serve coal mines. After the Civil War, Skip told us, "carpetbaggers" bought land for coal mining rights. Today, 70 percent of the natural resources of West Virginia are owned by outside corporations. "We were sold out ages ago and they're bleeding the area dry," Skip said. "The parts of West Virginia that today have poverty problems are the coal areas. Thriving, clean communities are not where there is coal."

The region had a long history of labor problems between miners and their boss companies. Skip told us about America's worst industrial disaster, "that almost no one knows about." The 3.7-mile long Hawks Nest Tunnel took three years to drill, beginning in 1929, to move water from behind a dam on the New River to a hydroelectric plant at the Gauley Junction. Miners for Union Carbide drilled through sandstone, but were given no dust control or breathing protection. Deadly silicosis killed thousands of them. The fate of the Depression era workers, desperate for work despite the deadly conditions, remained nearly invisible to most of the world. Many who died were buried in unmarked mass graves. A historical fiction novel, *The Hawk's Nest,* by Hubert Skidmore was published about the disaster in 1941, but the Union Carbide Company forced Doubleday to pull all copies from publication, helping to bury the grim story.

In recent decades, Mountaintop Removal (MTR) coal mining became the preferred coal-mining technique, a highly mechanized approach that employs fewer miners, so that coal industry employees in West Virginia dropped from 150,000 in 1950 to about 18,000 today. In the coming days, we were to learn much more about MTR coal mining's devastating impacts to rivers, soils, and vegetation, its threats to human health and property, and the political powers in West Virginia that defend the controversial practice, choosing short-term economic incentives over long-term societal and environmental health.

Driving from Charleston, the capital of West Virginia, to meet with Larry Gibson, we passed a billboard that read: "West Virginia Coal, the Real Power behind America." We were about to learn how powerful King Coal really is in the state. For 250 years, Larry's family has owned the summit of Kayford Mountain (37°58'N). He is known as "the Mountain Keeper," a staunch protector of his home and opponent of MTR coal mining, which turns mountains inside out, as coal companies blast them apart to get at thin seams of coal. The coal is separated and the rest of the mountaintop, now considered "debris," is shoved downhill into stream canyons. Over 500 mountaintops, 1.2 million acres, and more than 2,000 miles of streams have been destroyed. Flat terraces covered by unproductive rubble are left behind where mountains used to be.

We drove up a two-lane highway through a canyon where every few minutes a large truck full of coal went downhill. A small stream paralleling the road was red with acidic mine drainage. Larry's property on Kayford Mountain is a leafy island in a sea of mining operations. He made us welcome, but only after a wary greeting. His resistance had generated drive-by shootings and several violent incidents. He had a bulletproof vest hanging on a wall in his cabin, a gift from some supporters. He had never worn it, but kept it as a reminder to be cautious around strangers. He told us that the truck drivers had surely noticed and reported, on Citizen Band radios, our arrival that day in a Prius (not a common car in West Virginia).

In 1993, the Massey Coal Company made Larry and his family an offer for their land. "I got them angry when I wouldn't let them have it," he told us. "They said, 'We're going to take you off that mountain, give you some buying power. We're going to give you $140,000. How does that sound?' Like I should be flattered. They didn't understand that there's some things in this life..." He paused. "What do you have so close to your home circle of life, that's not for sale? Someone pushes on you and destroys it? How far are you going to go to stand for it and what would you do?"

Larry was 63 years old when we visited, a short, spry, muscular man with a white mustache. Though we were eager to see the MTR site that would become visible after a short walk through the trees, Larry first invited us into his simple cabin and made coffee. "My wife says I should serve toothpicks with my coffee," he joked, but it tasted fine.

Larry asked us to look at his walls, where he had framed copies of magazine articles and photographs of celebrity visitors to his hilltop. "Notice how everything's on with screws? When they're mining, dynamite goes off 10 to 15 times a day." Today, Larry must haul water to the cabin because blasting lowered the water table. Twenty-five tons of explosives are used each day across Appalachia.

Natural resources are destroyed by MTR, but also a way of life. At one time, 100,000 people lived in the canyon below Kayford Mountain, called Cabin Creek Hollow. As the mines bought people out, "They knocked over 20-something schools," Larry said. "We got one left. We had a school here, on top, and a church, a general store, about 60 families living right here. This land was everything. It was our doctor. Our food source. Our water source. Our medicine, shelter, warmth. Most people don't understand when I say things like, 'A mountain is a live vessel. A live body.'"

By the time he was five years old, he was helping in the farm fields, hoeing or pulling weeds at that young age. "We had a family coal mine here," Larry added. "Kids worked the mines here up to 1938, until the child labor law, sorting coal on a beltline. If you owned a drift like that, you worked it yourself."

The family moved off the mountain when Larry was 11 years old. After he retired from General Motors, in 1985, Larry moved back to the mountain. "It took me four years to clear plants that had covered the family cemetery," he said. "We had two family cemeteries, one on this side of the mountain and the other surrounded by coal. That one they destroyed."

Federal exemptions to the Clean Water Act were granted during the second Bush administration to allow MTR coal mines to push their debris into waterways, classifying the mine debris as "fill" rather than "waste," and eliminating requirements for buffer zones around streams. Those exemptions, so contrary to the purposes of the Act, should be enough to stop MTR valley fills, but the problems go further. Water pollution also results, because toxic chemicals leach downstream and the thousands of gallons of water used to wash each ton of coal during processing leave a toxic slurry that must be stored behind dams. Sludge reservoirs have become a looming threat to communities in the hollows downstream.

Larry said, "We've got young kids going to school beneath a preparation plant less than five miles from here as the crow flies, at Marsh Fork School. You've got one of the largest reservoirs in the world, called Shumate, 9.7 billion gallons. That dam is 89 feet taller than the one on the New River Gorge. It's not completely full yet. If it breaks, 17 miles away from the breaking point, the line of waste coming at the people will be 40 feet high. How many people you think will be dead by that time? We have 25 of these storage dams in this area."

The concern is very real. On February 26, 1972, a flood of waste sludge, cresting over 30 feet high, dropped onto 16 towns in Buffalo Creek Hollow (37°46'N). Within a few minutes, 125 people were dead, 1,121 injured, and over 4,000 left homeless; 561 residences and 30 businesses were destroyed. The Pittston Coal Company, trying to avoid liability, called the dam collapse "an act of God." In eastern Kentucky, we would soon visit a town hit by a coal waste flood that swept down 90 miles of river in October, 2000.

It seemed inexplicable that, with so few people actually employed in coal mining today, the majority of people, those who lost jobs and property, would not have a stronger voice in the region's politics, which currently favor the industry.

"For me, I've been fighting this so long that I think everybody has the sentiments that I have, but I've seen that people are intimidated," Larry said. "That's why the coal companies are able to do what they're doing. You make reference to democracy, something we're supposed to have. In this part of the country, we don't have it."

Larry has spoken at the United Nations. He handed out shirts from his organization, the Keeper of the Mountains Foundation, with the printed message: "We are the keepers of the mountains. Love them or leave them. Don't destroy them." There, he heard a debate about energy and "coal as a necessity, something we got to have," he said. "A speaker said, 'We can't handle any other industry in West Virginia." Now why do you think he said we can't do that? Because we have mountains."

Mountaintop Removal coal mining chewed away Kayford Mountain to the edge of Larry Gibson's family property

Walking out to the end of the ridge along a leafy path through an orchard, we realized there were no birds flying through the branches. "From 149 species of birds," Larry said, "we're down to 39. Every year I have people come in, checking our birds for us. I don't hear any birds."

At the crest of the hill, suddenly everything changed. As far as the eye could see, the mountains were gone, peeled away on three sides of Larry's land. We looked down on flattened terraces of rubble. At the distant edges, valley-fill debris created huge walls where streams used to run. Larry had tears in his eyes as he surveyed the scene, and so did we. The damage was so permanent and so short-sighted. We thought about how we would feel if our hometown mountains were blasted into rubble.

Out in the open, away from the trees, the wind suddenly

began gusting. The ground at our feet was rough cobbles, some baseball-sized pieces. ""It won't be the same in my lifetime, maybe forever," Larry said, kicking at the rubble in frustration. "What I'm showing you is what it will look like for many centuries to come. It takes about a thousand years for an inch of topsoil to form in these mountains." Green tinted hydroseeding helped provide a temporary illusion of green. "That's their 'restoration,'" Larry told us, "with an introduced Asian grass."

The MTR acreage totals spread across four states. As of 2008, Kentucky had 293 mountaintops removed totaling 574,000 acres; six mountains in Tennessee totaled 78,000 acres; Virginia's numbers were 156,000 acres on 67 mountains, and West Virginia had been most affected, with 352,000 acres on 135 mountains. (Geredieu, 2009).

When we visited Larry, the Environmental Protection Agency (EPA) had just decided to re-evaluate all Mountaintop Removal mining permits and to assess effects of MTR valley fills on aquatic ecosystems. The draft report that emerged concluded that streams are being permanently lost due to burial, chemical pollution is occurring downstream (including selenium concentrations elevated to toxic levels for fish and birds), and aquatic communities "are consistently and significantly degraded." (U.S. EPA, 2009). Before any findings were made, West Virginia politicians derided the decision to evaluate MTR impacts as, in itself, a threat to jobholders and the state's economy.

To us, re-evaluation sounded like an important first step.

"A good sign," Larry agreed, but he was impatient with the slow process: "I would like to see all coal mines stopped today and see them leave. If they left today, they wouldn't destroy anything else. I call for the abolishing, the abandonment of coal because it kills. Simple. It kills."

What we saw happening on Larry's mountain was reminiscent of techniques used in modern gold mining, which chew up hills, heap the material and leach out the gold with cyanide. It also reminded us of the late 19th century hydraulic gold mining in California, which used giant water cannons to wash hills down and pass the soil through sluice boxes to remove the gold, sending massive amounts of mud "wastes" downstream. Such willingness to demolish a living landscape for its hidden bits of gold was short-sighted, and hydraulic mining was eventually stopped in California because of its downstream impacts on cities and farms.

While coal currently is used to generate 45 percent of the electricity in the United States, Appalachian MTR coal provides, at most, 4.5% of the nation's electricity. We discussed energy alternatives. Windmill generators have been proposed to make electricity while helping the economy in West Virginia. "Some say windmills kill birds and bats and make a funny noise. Hell, this here," Larry said, pointing to the devastated scene in front of us, "kills everything. You don't like a windmill, you can take it down, but if you don't like a Mountaintop Removal mine, you can't put the mountain back."

Larry started telling people about Mountain Top Removal in 1986, figuring back then that once people saw what was happening, it would immediately stop. Twenty years later, after showing his place to more than 14,000 people, his dedication to the mountains in the face of intimidation from Big Coal and many personal threats, is truly an inspiration. Yet, MTR continues, and that is heartbreaking.

"In the most powerful country in the world, so they say, the most educated country in the world, so they say, coal was once essential because we were a young country; like a child," Larry told us. "Now we're walking upright. Now we're powerful and educated. Why are we still using the most barbaric form of energy known to mankind?" He answered his own question: "It only takes two-and-a-half minutes in a powerplant to burn a ton of coal that sells for $245. Think how much money they're making. It's not about energy or jobs. It's about making a profit."

The last thing Larry told us, on that day at Kayford Mountain, was that a senior citizen march protesting MTR coal mining would begin at the state capitol the next morning. Larry was going to join the group and invited us to come along on that first morning of the five-day walk.

The West Virginia State Capitol dome gleamed gold in the background as leaders from Climate Ground Zero, Mountain Justice and the Mountain Keepers Foundation addressed the gathering walkers.

Coal was in the newspaper headlines that morning. West Virginia Senator Byrd had surprised the coal industry by taking Massey Energy to task for refusing to help pay to move the Marsh Fork Elementary School, the one Larry had described that was threatened by a coal slurry reservoir above the school. (Ward, Jr., 2009).

Organizer Roland Micklem, 81 years old, described the message he hoped the group would deliver in the coming days: "Coal companies do not own the mountains; God owns the mountains." He planned to seek an audience with Massey authorities at the end of the march. The group expected to face harassment, particularly as they approached the mine.

Some of the group carried signs and banners that read: "Clean Energy Now;" "The Mountains Belong to God, Not King Coal;" "Elders Say No to MTR;" and "How Much did King Coal Pay for Your Soul?" As they walked along, there were occasional honks of support from passing cars. Roland, leaning on a walking stick, told us he felt compelled to take this action. "I'm accountable to myself, and I don't want to look in the mirror, to regard myself, and feel I didn't do everything I possibly could to stop this Mountaintop Removal."

One of the walkers that morning was Jesse Johnson who had twice been a candidate for governor for the Mountain Party of West Virginia. He was not currently running for anything, but explained why he was there and the motivations behind his political career: "You used to be able to sit at my farm in Lincoln County and not hear anything except nature. Now, I'm hearing blasting going off. The largest MTR site in the world is coming closer all the time. In the future, I can see myself standing on my great-great grandparents' grave, where I intend to be buried and trying to protect it like Larry." His political action career began because every household near his home was hit by high cancer rates because they were using well water contaminated by an industrial dioxin dump.

Jesse addressed the question we had posed to Larry: with so many jobs lost to modernization in coal mining, why do those hundreds of thousands of former workers and their families not support a different approach to the local economy? "The problem is," he said, "they get hornswaggled into believing the myth [that coal mining is good for the region]. The extractive and polluting industries are where all the dollars are, and they purchase more than 75 percent of the advertising in the media, and have come in now and bought the media."

That answer made us recall the billboard message: "Coal, the Real Power Behind America." Coal culture runs deep and wide in Appalachia. Generations have worked in the mines. One local told us they either mined coal or flipped hamburgers; there was nothing else. Mining companies have long been paternalistic, providing housing, health care, and a predictable paycheck, but loyalty to coal carries a steep price. Hopefully, in the future, there can be new opportunities for families to earn sustainable livings. Hopefully that will happen before all the mountain tops are blown away and shoved into the river channels.

West Virginia's state motto is "Wild and Wonderful," but coal's impacts on people, land, water, and the atmosphere turn that slogan into a sad irony. Yet, much beauty remains that is worth saving, and many West Virginians we met had not given up hope.

In January 2011, the EPA revoked an Army Corp of Engineers permit that had been issued in 2007 for a large MTR mine in West Virginia. Spruce No. 1 Mine, owned by Arch Coal, had planned to clear 2,278 acres and dump material into six miles of streams near Logan City (37°51'N). The EPA announcement read: "The proposed Spruce No. 1 Mine would use destructive and unsustainable mining practices that jeopardize the health of Appalachian communities and clean water on which they depend. We have responsibility under the law to protect water quality and safeguard the people who rely on clean water." (Malloy, 2011).

UPDATE 2020

In 2018, the Trump administration rolled back the Obama era 2016 Stream Protection Rule that required MTR mining companies to monitor and restore streams where mine wastes had been deposited. The EPA rule had been enacted after 8 years of study and 94,000 public comments. The newly elected administration also defunded a National Academy of Sciences study to document the human and environmental effects of mountaintop removal coal mining.

Mid-western Rivers and the Population Center of the U.S.

Coal was not yet done with us as we entered eastern Kentucky. One of the worst coal industry disasters occurred on October 11, 2000, when a sludge dam operated by Massey Energy gave way and sent 300 million gallons (25 times the size of the Exxon Valdez oil spill) of toxic, black goo downstream through the community of Inez (37°52'N). The sludge moved down Coldwater and Wolf Creeks to the Tug Fork and some material reached the Ohio River. No human lives were lost, but the EPA called it "the largest environmental catastrophe in the history of the southeastern United States."

In Inez, they were preparing for a Harvest Festival when we arrived. A young woman selling popcorn had been a high school cheerleader when the flood of coal waste came and remembered, "It was thick, black, and gooey, so solid that thrown rocks bounced off of it. They brought in big vacuum machines to clean it up. Things are clean now and they widened the creek in case it ever happens again." She finished that thought with a giggle. The idea of it happening again must be unnerving, yet the townsfolk remain.

Through West Virginia, across Kentucky, and parts of Indiana and Illinois, we explored a slice of the vast Ohio River watershed: the New and Gauley Rivers, the Kentucky River, and the main course of the Ohio River en-route to the Mississippi. Settlers, settlements, transport and the rivers' inclinations to flood were elements in that story.

156

The look of the forested land changed as we moved westward through Kentucky to Fort Boonesborough State Park (37°56'N). There, beside the Kentucky River, a party of settlers led by Daniel Boone arrived on April 1, 1775. Seeing the advantages of the river, they established Kentucky's second pioneer settlement. The riverside setting has changed since Boone's day, because of locks built early in the century to facilitate barge travel.

Near Port Royal, Kentucky, we had lunch with author and poet Wendell Berry, his wife, Tanya, and their grown son, Dave, at their farm. The Kentucky River is visible from their house and is part of the special character of the property. "It lies within a hundred steps of my door," Wendell wrote in an essay titled "A Native Hill." (Berry, 2002, 3). Small streams on his property feed the river – which merges with the Ohio River just a few miles downstream – and water figures in many of Wendell's poems and essays. He wrote: "The creation is musical, and this is part of its music, as birdsong is, or the words of poets. The music of the streams is the music of the shaping of the earth, by which the rocks are pushed and shifted downward toward the level of the sea." (2002, 19).

Wendell is tall and lanky with penetrating blue eyes. Tanya was warmly welcoming and an excellent cook; she grew up in Mill Valley, California, another 38th parallel location and speculated that the similar angle of light might be something agreeably familiar about Port Royal. Wendell became most animated as he drove us around the farm and through his "biscuit woods," a mix of black walnut, butternut, ash, and tulip poplar trees that provide good fuel for baking. It was wonderful to spend time with someone whose books and poems so eloquently explore the importance of place and home and sustainable farming.

Lexington bills itself as "The Horse Capital of the World" and, besides its famous horse stables and racetracks, houses the International Museum of the Horse (38°08'N). Though our primary interest along the 38th parallel was water-related environmental topics, an intriguing horse theme had become evident. In China, there were the ancestors of the modern horse, the Przewalski's breed. Turkmenistan's Akhal-Teke horses are ancestors of the Arabian breed. Spain's Andalusian horses originated along the line, and the Atlantic Coast of the

United States had the island-dwelling Chincoteague ponies. Spanish mustangs run wild just east of our Mono Lake basin home.

Certainly, different breeds of horses are associated with other latitudes, but a connection between 38°N and grazing animals of all types is more than coincidence. An article in the *Stockman Grass Farmer* advised ranchers who want to produce grass-fed beef to move close to the 38th parallel: "Almost all of the year-around grass finishing in the world occurs within 100 miles north and south of this degree of latitude. It is far enough North to grow cool-season perennials but far enough South to avoid long periods of deep snow." (Nation, 2010, 24)

At Louisville, the state's largest city, an interstate freeway was built between the city and its Ohio River waterfront, blocking views and access (38°15'N). The freeway remains, but now a lush park, with sculptures and architecture that celebrate water, draws people under the freeway to reach the river's edge and views of its bustling river traffic. The city of

A tugboat and barge on the Ohio River at Louisville, Kentucky

San Francisco went through a similar recovery of its bayshore waterfront after an earthquake made it feasible to correct a badly located freeway. Out in the river opposite Louisville are the Falls of the Ohio, where a series of rapids interfered with river traffic until the government built a dam and locks in 1830.

On the Indiana side of the river, one family has run the Schimpff's Candy Store, in Jeffersonville, since the 1850s. Red marks painted on the outside storefront mark where the flood of 1937 reached, when over a thousand miles of the Ohio and Mississippi Rivers left their banks. It was the worst natural disaster in the United States before Hurricane Katrina. Over 800,000 people were displaced from flooded homes and hundreds died.

A chilly morning brought our first view of the grand Mississippi River when we arrived at Grand Tower, Illinois (37°37'N). Looking for a place to warm up, we found the Mississippi River Museum being used as an office by the County Clerk, Charles Burdick, who had navigated river boats on 12,350 miles of inland navigable waterways as a pilot and captain. The Mississippi River extends from Minnesota to New Orleans, but its vast watershed feeds into the big river through major tributaries, like the Ohio and Missouri Rivers, and many smaller branches that reach into neighboring states, with connections to the Great Lakes, and, via the Gulf Intracoastal Waterway, to Texas.

Kaskaskia Island was visible on the far side of the Mississippi from Fort Kaskaskia, the only piece of Illinois *west* of the river (37°58'N). The land ended up on the wrong side after the Mississippi River changed course in an 1881 flood. "Kaskaskia, where the West began, Mother of 1,000 Cities, Paris of the West," had been the first capital of the Illinois Territory and capital of the state until 1820, but the flood washed away its physical connection to the rest of Illinois. Our visit to Kaskaskia found only quiet farm fields.

A highway bridge at Chester, Illinois, led us to the beautiful little French-flavored town of St. Genevieve, the oldest town in Missouri. A car ferry also crosses the river there, at *exactly* 38 degrees. Of course, a ride over and back as walk-on passengers was irresistible. The Mississippi was perhaps a mile wide there and looked exceptionally swift and powerful from a small ferry boat chugging back and forth across its surface.

"I'm out planting a forest. Please leave your name and number and I'll try to get back to you before it matures." That is the answering machine message heard in the opening scene of a video produced by the Pioneer Forest about their founder, Leo Drey.

Author Wendell Berry had suggested a visit to to learn about the Pioneer Forest's half-century long experiment in sustainable forestry near Salem, Missouri (37°34'N). Forest Manager Terry Cunningham and two other foresters told us about the unique forest management. The vision began with Drey, who began purchasing land in 1954, eventually totaling 150,000 acres of forest lands, most from a distillery company that had used the oak wood for barrels. It became the largest private landholding in Missouri, land that remains open for public use. Drey aimed to demonstrate methods of sustainable logging of the oak forests, and to avoid going broke in the process.

The Pioneer Forest follows a long-term cycle of single-tree harvesting that makes for an aesthetically pleasing forest with trees of different ages, minimizes erosion and run-off on the watershed, and improves reproduction and overall health of the trees. The best trees are left in place, instead of the worst, and the values and character of the forest remain. This approach is in contrast to clear-cutting favored by the Forest Service in such mixed hardwood forests, where regrowth produces thickets that must be aggressively thinned. An aerial photograph of some of their forest revealed widely spaced stumps within an otherwise untouched-canopy of large trees.

Minimizing watershed impacts while harvesting trees has been one of the Pioneer Forest's significant contributions. The Forest staff have shared their system with other private forest landowners, and also hope to influence policies in the Ozark National Riverways Park, where the Current River borders their land holdings. The park, part of the National Park System, owns only a narrow quarter-mile strip of shoreline along the riverbanks. Canoeing and fishing recreation are no problem, but erosion problems created by all-terrain vehicles are, as they are often driven into the river and on its erosion prone banks.   The Pioneer Forest staff hope for management more in line with standard national park service resource protection.

In nearby Edgar Springs, Missouri, we located an intriguing geographical landmark.   Each national census identifies the "mean center of population" of the nation, a rather strange statistic that calculates the point where an imaginary, rigid, flat map would balance if all residents of the nation were of identical weight.   When Alaska and Hawaii became states in 1960, the center point moved about two miles farther south and ten miles farther west.   For the last several decades, the point has hovered around the 38th parallel, gradually shifting westward and southward.   The 2000 Census placed it 2.8 miles east of Edgar Springs, a tiny town in central Missouri (37°41'N; 91°48'W).   Its distinction was noted on a highway sign, but we went into town to see what else might be found.

The lady at the general store said she thought maybe there was a stone marker at the north end of town with information about the designation, but added, "I never go that way."   We found the marker outside the gates of a cemetery, where other stones told another local population story.   (The 2010 census has since documented a continued southwest drift, new "population center" markers will be needed near Plato, Missouri, at 37°30'N).

Oak woodlands gave way to prairie, as rain chased us the rest of the way across Missouri, out of the rolling hills onto the plains of Kansas.

UPDATE 2020
In the spring of 2020, for the second year in a row, the region where the Mississippi River and Ohio River meet braced for multi-billion-dollar flood losses.   Rivers were brimming and farm fields were too soggy for planting. Such persistent wet conditions were consistent with climate change predictions for that region.

Seagulls in Kansas and the Santa Fe Trail

The Kansas prairie is, despite first impressions, not perfectly flat, but features low, undulating hills. Prairie grasses, gone dry in the autumn, replaced the green of the Midwest heartland. Tallgrass Prairie National Preserve was established by Congress in 1996 in the Flint Hills of eastern Kansas (38°26'N). Because of local opposition to constraints that typically come with a national park, legislation establishing this preserve limited National Park Service acreage to 100 acres; the Nature Conservancy owns the rest of the 11,000 acres.

Bundled up against cold wind and occasional raindrops, we walked away from the ranch buildings that serve as visitor center and park headquarters. Once the trail dropped down a hill so that no structures were visible, the open sky and expanse of unfamiliar vegetation took charge of our awareness. Tallgrass prairie plants are unfamiliar because less than four percent remains of the ecosystem that once covered 140 million acres in North America. The Preserve protects a significant remnant, although Bison (*Bos bison*), the keystone species on the prairie, were missing.

Within a week of our visit, that lack would be resolved by moving 13 Bison from Wind Cave National Park in South Dakota. The following spring, the first Bison calf in 140 years was born on that remnant of tallgrass prairie landscape. Two more births followed. Local cattle ranchers had voiced concerns that disease that might be passed to their herds by Bison, but other parks, like Wind Cave and Yellowstone, had demonstrated how to manage herds to avoid problems.

Re-introducing Bison was only one of many contentious issues before the Preserve's creation, described by William Least Heat Moon in his book *PrairyErth*. The book focused on Chase County, Kansas, because that author wanted to explore the "center" of the United States, located by tracing diagonal lines from the four corners of the lower 48 states – a calculation that also coincides with the 38th parallel.

"One would as soon expect to find sea-gulls in Kansas," Mark Twain wrote after visiting Mono Lake." (Twain, 1873, 268). We discovered there *are* seagulls in Kansas, lots of them. Also cormorants, pelicans and phalaropes. Central Kansas has the largest inland wetlands in the country. It was a treat to cross miles of prairie and arrive at Quivira National Wildlife Refuge on the sunny day they celebrated National Refuge Week (38°12'N). Blue skies had returned and many familiar birds were out on the slightly saline water. Quivira and neighboring Cheyenne Bottoms provide critical habitat for migrating birds on large expanses of surface water in the prairie landscape.

Biologist Rachel Laubhan and Ranger Barry Jones welcomed us to their event, shared much information and fed us a roasted pig dinner. At sunset, at the northern edge of Quivira, flocks of White-fronted Geese (*Anser albifrons*) flew in for the night, ducks were settling, blackbirds massing, and all were honking, quacking, or singing.

The next morning, we stopped at the new Kansas Wetlands Education Center at Cheyenne Bottoms, a showplace for environmental education and a joint effort between Fort Hays State University and the Kansas Department of Wildlife and Parks. A Western Hemisphere Shorebird Reserve Network sign brought back memories of installing the same signs years ago at Mono Lake.

There is competition for a limited water resource between the wetlands and neighboring farms in that watershed, called the Rattlesnake Creek basin. On the portion of Cheyenne Bottoms managed by the Nature Conservancy, a legal battle over protection of public trust values was fought by the Conservancy with local farmers, to ensure adequate water reached the wetlands. To avoid that kind of lawsuit, the U.S. Fish & Wildlife Service instead negotiated a collaborative "Rattlesnake Creek Partnership" intended to keep refuge lands

Gulls and other migratory birds thrive in central Kansas at the Cheyenne Bottoms National Wildlife Refuge

saturated, without putting farmers out of business. Mark Sexson, a Public Lands Supervisor with the Kansas Department of Wildlife & Parks, explained that, instead of cutting a farmer down from 16 inches of water a year to only five inches, "which no one can use to raise anything, they're given 25 inches to use over five years. So if they want a high water-use corn crop, they may use 16 inches of that 25 in one year. Then they may decide to not irrigate for several years. Or they can grow wheat, which doesn't take very much water, and use five inches a year to irrigate for five years."

It sounded like the well-intentioned collaboration was floundering at that time, but the objective to stretch limited water resources between agriculture and the wetlands environment seemed worth pursuing.

Heading out of Great Bend, Kansas (38°21'N), we were following the Arkansas River across the plains of central and western Kansas along the 38th parallel, all the way to eastern Colorado. The historic Santa Fe Trail closely paralleled the Arkansas River (pronounced "Ar-Kansas" locally) – the most reliable surface water in that region. From 1822 to 1880, traders and settlers used the trail to connect Missouri to New Mexico, with a mountain branch leading into Colorado.

164

Pawnee Rock was a major landmark for Santa Fe Trail travelers because it rose above the flat prairie and marked the approximate half-way point to New Mexico. Hundreds of pioneers carved their names in the sandstone. Though much of the rock was quarried in the late 1800s, stairs to an elevated platform gave us a long, unbroken view across the prairie from its former height.

Not far away was Fort Larned, on the Pawnee River fork of the Arkansas River. In 1859, 2,300 people, 1,970 wagons, 840 horses, 4,000 mules, 15,000 oxen, 72 carriages, and over 1,900 tons of freight left Missouri for New Mexico. Because livestock and freight were being lost to Indian raids, the fort was authorized and two companies of U.S. Infantry arrived there on October 22, 1859. (Tiller & Toiler, 2009).

Gigantic cattle feedlots surround the famous Queen of the Cowtowns, Dodge City, where statues of Wyatt Earp and a longhorn steer celebrate the era of cattle drives from Texas to the railhead by thirsty, rowdy cowboys, who were themselves corralled by gun-toting lawmen. Just east of town we had crossed the 100th Meridian – the 100°W latitude line (at 37°48'N). Beyond there, as we continued westward, we would increasingly appreciate the truth in John Wesley Powell's description of that "line between the humid east and the arid west," marking where less than 20 inches of rain fall each year. In that region, Powell wrote,

"agriculturalists will early resort to irrigation." (Powell, 1879, 2-4). Suddenly, hair that curled in Eastern humidity fell straight again and lips felt dry.

Garden City, Kansas is a short drive west of Dodge and home to the 3,670 acre Sandsage Bison Range, where State Park Manager Tom Norman showed us the herd of 90 Bison about to be reduced to 60 in an annual round-up. Their shaggy bulk reminded us of the Yaks of western China (they are in the same genus and Canadian ranchers have interbred them to produce "Yakalo"). Bison have been raised here ever since "Buffalo Jones," one of the founders of Garden City, rescued a few of the last remaining animals from extinction. Offspring from 57 calves he brought to Garden City make up the herd south of town, and became the source of new Bison herds around the world.

Near Dodge City, Kansas, the "Queen of the Cow Towns, sprawling feedlots handle thousands of cattle fed with grain grown using the overdrafted Ogallala aquifer water

A few scattered juniper trees on the preserve were evidence of a quixotic attempt to create forests on this sandsage prairie. The hope was that planted trees would modify the climate, actually increase rainfall, and make conditions suitable for a sustaining forest. The 30,000-acre Garden City Forest Reserve was established in 1905 with experimental tree plantings of juniper, mulberry, and honey and black locusts. In 1908, another 272,387 acres were added, extending west all the way to the Kansas-Colorado border. With that addition, the Reserve became the Kansas National Forest. By 1915, millions of tree seedlings had been planted, but climate and soil realities doomed the project.

While driving through the pastures, Tom pointed out the cooling towers of a coal power plant visible in the distance. Needing lots of water, the power company had purchased 20,000 acres with water rights formerly used for crop irrigation. Controversy ensued, but, "To me, and most of us in the wildlife field out here," Tom said, "we thought that was the most positive thing to happen in southwest Kansas for wildlife in 50 years. I was pretty excited about seeing so many [irrigated crop] circles going back to permanent cover. One of these days, it's not going to be economical to pump water and put it on crops anymore. What's going to happen when it isn't? Are they just going to turn off the wells and back away?" Tom was referring to the groundwater pumping issue that dominates the environmental and economic future of that region: overdrafting of the Ogallala Aquifer.

UPDATE 2020
The Kansas Wetland Education Center (KWEC) at Cheyenne Bottoms that was brand new the day we peeked in its windows, now has become a dynamic environmental education facility serving school groups and the public. KWEC completed an exhibit enhancement project in November 2018, adding new exhibits with interactive features. New exhibits include a one-of-a-kind wetland floor projection game, augmented reality watershed simulator, wetland children's activity area, "build a bird" and "build a plant" stations, shorebird and wetland plant adaptation magnetic boards, aquatic invertebrate viewer, and wetland critter doors.

Mining the Ogallala Aquifer

The Ogallala Aquifer extends from Nebraska down to Texas. This enormous source of deep groundwater allows the region to grow corn and grain and raise most of the nation's cattle on high plains that only became very productive after efficient groundwater pumps and center-pivot irrigation sprinklers were invented.

At the Kansas Groundwater Management District #3 in Garden City, Executive Director Mark Rude, dressed casually in blue jeans and a plaid shirt, somehow stays upbeat as he faces undeniable facts: there is less than an inch of recharge into the aquifer each year, but in places withdrawal is happening hundreds of times faster than that. Many pumps now work at reduced capacities. When the water runs out, the economy and environment of this vast region must change.

Meanwhile, in the attempt to slow, though not yet control the situation, each water user in the District is limited to a set allotment that is enforced with well-monitoring meters and sensors. The gloomy news was that the groundwater depletion had not been stabilized, because the limits on pumping were not adequate. "Economists are winning; not hydrologists," Mark told us.

By drilling test wells, early ranchers became certain there was a vast, "inexhaustible" underflow, but most of it was down too far for early pumps. Garden City got its name because its windmills tapped into a fairly shallow water table to irrigate gardens and fill surface ponds. "When deep turbine pumps came in," Mark said, that opened up the range land; you could lift from any depth." Local wells reach down 60 to

70 feet, but most wells tapping the Ogallala aquifer go down several hundred feet, and farther south, along the Oklahoma border, some have chased the declining water table down 700 feet. The cost of energy to lift water from those depths becomes a limiting factor.

Center-pivot sprinklers were the other key innovation. An aerial photograph of the groundwater district, mounted on the wall of the conference room, showed mile after mile of green circles fed by 10,000 wells. "The center-pivots allowed massive development in the 1960s and into the '70s," Mark said. "Wells could be put right beside the sprinkler pivots. You could take $50-an-acre ranch ground and make it $1000-an-acre ground." A state engineer during that era approved pump permits without regard for sustainable yield. Saying "no" to new wells, when neighboring ranchers were pumping, was politically difficult. The easy choice led to too many pumps tapping the aquifer.

We asked Mark to predict, with the trends he saw, what it will be like in western Kansas in the future. "I'm an eternal optimist," he said. "I think conservation of the water supply will increase. There will be a lot more managed recharge, a lot more accounting for every drop." But then he described the problem between hydrologists and economists, adding, "In Eastern Kansas, with more water, they can manage for sustainability, but in western Kansas, our long term plan is to deplete the aquifer."

Mark knew about the decision to protect Mono Lake by reducing stream diversions, which was based on the public trust doctrine. "But that likely wouldn't have happened if the water was below ground instead of on the surface," he said. Effects on an underground aquifer are out of sight, and the classic "out of mind" result is a big problem.

"We have 10,000 flow meters out there on wells. Sometimes those meters seem to break down at the most inopportune time from a compliance standpoint," Mark said, wryly. "Our answer to that problem is a temperature logger, an inexpensive device to strap onto a pipe that records when the well is operated, because groundwater is cold."

To show us how the monitoring works, Jason Norquist, assistant manager, took us out to watch as he installed

169

Jason Norquist checks groundwater well monitors to see that farms comply with pumping limits from the Ogallala aquifer in southwest Kansas

temperature sensors on several wells. As he drove, he explained how he knew when one farmer had been playing games with his meters: "Meters were running backwards because he had them installed backwards. The patterns where there are leaks, of algae and moss on the pipes, didn't match up." Washers and the nuts had lost their rust when a wrench was used on them. Jason found five more wells at the same ranch with meters on backwards, so he called the farmer up and told him what he had found. They went over that farmer's pumping hours and, since then the farmer's attitude had turned around.

"Most of these guys," Jason added, "keep track of new innovations and are bringing in drip irrigation systems and more efficient ways to use the water. One of the first in the area to put in subsurface drip took his water use from 25 inches per acre, down to about 14 or 15 inches. Then his dad and his grandpa...well, they grew up the old way. They like to *see* the water. With that subsurface drip they can't see any, so whenever he shut his well off, they'd come out and turn it back on. There's a generation gap."

As Jason installed a temperature sensor on a pump in a field of corn stubble, he explained that the old practice of flood irrigation extended across entire 160-acre quarter-sections, to the corners. With center-pivot sprinklers, the resulting circles reach only about 125 acres. Drip irrigation is more efficient per acre, but also allows farmers to plant out to those corners again, so water savings, overall, are not as high per quarter-section as might have been hoped from the more efficient system.

That sounded like California's Central Valley Project, which had a goal to solve groundwater overdrafting with imported water that was not achieved, partly because the new water was instead used to extend irrigation onto new acreage.

At breakfast, we spoke with Mark Sexson, the Regional Public Land Supervisor with the Kansas Department of Wildlife & Parks. His responsibilities extend to just about anything to do with water in western Kansas. Mark confirmed many details we had been hearing about the Ogallala aquifer overdrafting issue. He pointed out that "use it or lose it" water rights laws perpetuate a lack of incentive to conserve. When we asked what he saw coming in the future for that region, Mark said: "As water declines in the aquifer, rather than just shut off water to people that had water rights...that's harsh...instead we should pay to retire the rights. Then put the land back into grass to perpetuate prairie chickens. As far as wildlife goes, the best thing that can happen is to run out of water fairly quickly," so that the wildlife habitat in the sandhill prairie can be restored.

Recharge for the aquifer in Kansas originates uphill in Colorado. The State of Kansas took Colorado to court over holding back too much water and won. Arkansas River water stored in an eastern Colorado reservoir now must be made available to Kansas. Mark feels that a part of society's problem is the "disconnect, whether it is wildlife or farm or water issues, between the urban vote, the people that have the power in legislatures passing laws, and the reality of life in the rural areas, where food comes from." Environmental education was important, we suggested.

Mark agreed. "My wife, Kathy, directs the Center for Conservation in Garden City. She gets 100,000 visitors a year, but even in Garden City they don't understand where their water comes from."

As Mark suggested, urban demand was another part of this Kansas water story. We met with Garden City Mayor Nancy Harness in a downtown coffee shop. She is an impressive leader of a cosmopolitan community, and began by talking about those problems of perception and long-range planning.

"There's a mindset that water is here to be used and acceptance that someday it will be gone," Nancy said. "In this part of the country, things seem so temporal. You can drive around here and see these little communities with remnants of homesteads, houses and churches. Our culture is 'use it and move on.'"

When that cultural pattern dominates, it becomes nearly impossible to effectively address long-term water supply issues. Our global travels had shown us, time and again, how strong connections to a place provide the motivation for sustainable lifestyles and successful resistance to short-term extractive policies driven by impersonal corporations or distant politicians.

As for water awareness and conservation, Nancy said, "We're too green." She did not mean "green" as in "environmentalism," but the color itself. "Everybody loves to water their lawns, and they water them, the sidewalk, and the street. Which is maddening because it is a finite resource. We [the city] did adjust the pricing so the more you use, the more you pay, trying to make average water bills reasonable. But we have yet to get truly serious about water. No one has ever spoken to..."

"Limits?" we suggested.

"Yeah. We're just in complete denial. There is no doubt about it. [People feel] you can turn it on, it's always there."

She told us about a neighbor whose sprinklers ran every day, rain or shine. We mentioned that "smart sprinkler" systems are available now that monitor moisture and decide whether to come on or not. Nancy laughed. "I don't think it's the sprinkler's fault."

The biggest employer in the Garden City region is the Tyson Foods Packing Plant. Earlier that day, Nancy had been on a tour of the cattle processing facility where 3,100 people work. "They kill 400 cows an hour; thousands of cattle every day, and contribute millions of dollars every month to our economy. It's huge. The cattle all come from within 100 miles of the plant. All the corn you see around here fattens the cattle at the feedlots, which then go to the packing plant. Hundreds of gallons of water are used to process each carcass." To secure the water they need, Tyson purchased thousands of acres with water rights (as the local coal power plant had done). Those acres also gave them a place to send their reclaimed wastewater, which fertilizes crops.

The plant attracts economic refugees to Garden City, because, until recently, they could always find low-skilled work. "There are 27 languages spoken by people at the plant," Nancy said. "It is the "United Nations." Folks can come here, earn $12 an hour and not have to speak English. But they're awful jobs. They come from Somalia, Burma..." She paused. "...and Michigan," she finished, laughing. "We had 25 teachers who came here from Michigan this year. I spoke to some of them and told them 'I'm sorry things are not good in Michigan and I'm glad you're here. I promise you, in February, *you'll* be glad you're here, when it's cold and snowy up north.'" Nancy welcomes this mix of cultures as a positive thing for her community, though it seems likely that the influx of newcomers is amplifying the long-term commitment issues in that locality.

Parallels between the dilemmas of unsustainable pumping from the Ogallala aquifer and Mountaintop Removal coal mining are striking: a short-term economic choice will leave future generations with enormous problems and environmental loss. It was another hard story to hear and another motivation for spreading awareness about the water sources to grow food and send the life-giving elixir to our taps.

The Ogallala groundwater basin extends into neighboring Colorado. We noted the changing appearance of the Arkansas River as we drove: wet with emergent groundwater in Great Bend, bone-dry in Garden City, but perking up as we followed the mountain route of the Santa Fe Trail across the state line.

UPDATE 2020: Ogallala aquifer declines are still being recorded, but have been less than a foot each year since 2015.

173

Colorado: The Headwaters State

The Arkansas River's headwaters are in Colorado, named "the Headwaters State" because so many rivers originate from the state's mountains. Water from Colorado reaches 18 states and Mexico. Snow that melts west of the Continental Divide, after passing through dams and aqueducts, may emerge from faucets in Los Angeles.

The Southern Rocky Mountains divide into a complex series of ranges with 14,000-foot peaks that made it impossible to stick closely to the 38th parallel as we headed west. In the following days we swooped north and south and north again through the Sangre de Cristo and San Juan Mountains, seeing some of Colorado's most scenic canyons, spectacular passes, and historic mining towns.

West of the city of Pueblo, the Arkansas River canyon climbs upward and the river becomes a swift mountain stream. We reached Salida, Colorado, 7,000 feet above sea level, on a night of early winter weather, and took a day off from traveling, hosted by former Mono Lake neighbors. There we learned about a battle being fought by local citizens against Nestlé Waters' plan to export local water to a Denver bottling plant. The concerned citizens we met that day were still reeling from a decision reached by their Chafee County Commissioners a few weeks earlier. Land-use and pipeline permits had been approved, allowing Nestlé to move 65 million gallons of water a year to an Arrowhead brand bottling plant in Denver. Twenty-five trucks would make the 240-mile round trip each day.

We met Vicki Klein and Betsy Steinberg in a cozy coffee shop near the river as rain and snow kept falling, and later joined John and Ann Graham, who are officers in the Chafee County Citizens for Sustainability group that led a campaign against the bottled water proposal. Their opposition was partly based on the number of trucks added to local highway traffic, but also the illogical waste generated by bottled water. "Taking the water doesn't provide local jobs," Betsy said. "They aren't even gassing their trucks up here." The county will receive about $80,000 a year, "but that's not much money considering the problems this is going to generate. They treated us like we're a bunch of stupid locals, asking, 'What if we put on the bottle label that it is from Chafee County? Would you like that?' It was insulting."

The "carbon footprint" of bottled water is amazingly high. In the United States, 28 million bottles of water are sold each year in plastic containers that require 17 million barrels of oil. The manufacturing process consumes three times as much water than what is inside each bottle. Text on a metal bottle made for the Citizens for Sustainability reads: "I bottle my own water."

John's hope was that "the public will just open their eyes and close their pocketbooks [to] companies like Nestlé. Bottled water is passé; bottled water is foolish; bottled water is 'not cool.' It is a last resort for emergencies only."

We were ready for a more positive story, and found one in the San Luis Valley after driving across a snowy pass that brought us back to the 38th parallel near the dramatic Great Sand Dunes National Park. The former National Monument's status had been upgraded to "National Park" after the Nature Conservancy purchased local ranches, foiling a water export scheme meant to serve development along Colorado's front range, while also creating a new National Wildlife Refuge. The "front range" is what Coloradans call the land east of the Rockies, where Denver and its thirsty suburbs are located.

The background to the San Luis Valley water story was given us by Paul Robertson, a white-bearded local project coordinator with The Nature Conservancy (TNC). We met at the Scenic Country Store Cafe, where the cinnamon rolls were a day old, but the waitress promised they would be good when warmed up with butter, adding, "If you don't like it, I'll feed it to the chickens." Coffee money, whatever amount one felt like paying, went into a donation can.

The ancient San Luis Valley is the largest alpine valley in the United States with over 5,000 square miles of valley floor; the State of Connecticut could fit comfortably within it. The valley was once covered by a hundred feet of water until, during a particularly wet era, water broke out to the south and formed the Rio Grande. Beneath sediments from the San Juan and Sangre de Cristo Mountains lay two aquifers, a deep confined one that generated "head pressure" to push artesian springs to the surface, and above that, separated by a clay layer, a shallow aquifer.

In 1835, the first diversion ditch tapped into the Rio Grande to irrigate farms. In time, enough water was taken to dry up the river. Ranchers started drilling wells in the 1930s, a process that continued until just recently. Wells now serve 2,500 center pivot sprinklers for irrigating crops. Most tap the shallow, unconfined aquifer, but enough extend down to the lower aquifer to reduce the pressure and stop the natural artesian flows that people relied on in the early years. Today, 40,000 people live in the valley.

In the early 1970s, the American Water Development, Inc. (AWDI) owned the Baca Ranch on the east side of the valley. As their name took no pains to hide, "they were really most interested in the water, to move it to the front range," Paul said. "At the time, the state didn't control groundwater; almost all Colorado water law was based on surface water." (In California, groundwater is *still* not regulated by the state.)

AWDI sold to American Stockman, backed by investment money from outside the region. "Money kept the battle going," Paul said. "One very flamboyant local guy was leading the effort to export water. We identified this valley as having high conservation values, both biological and natural resource values: the sand dunes, the water, the aquifers. Most ranchers and the land management agencies were very much behind us, because they were seeing water tables drop and springs go dry."

In 1999, the TNC bought the 100,000-acre Mendono-Zapata Ranch, the biggest purchase deal they had ever made. The Clinton administration supported the plan to upgrade to a National Park and also create a 60,000-acre National Wildlife Refuge. TNC made the purchase and then sold to the federal government, so that the former national monument, which had about 40,000 acres, now met the acreage criteria for a Great Sand Dunes National Park designation.

Center-pivot sprinklers irrigate circular fields in Colorado's San Luis Valley, in the watershed of the Rio Grande

"Everyone got what they wanted, except the people who wanted to sell water to the front range," Paul said. "That one stockholder made an interesting argument. He asked why it was OK to put water into potatoes and ship them out of the valley, but not OK for him to pump the water and ship it directly? Of course, if that had happened," Paul added, "he would have been a millionaire and no one else would have gotten a nickel out of it. The other way the money is scattered around the valley."

In recent years, Colorado has tightened up laws controlling groundwater. No new permits are issued now except for domestic use. There are 30 monitoring wells in the National Park and hundreds across the valley; if the water table drops too low, pumping must stop. "That happened in the 2002 drought," Paul said. "To maintain the aquifer, they closed down one thousand wells. Two years ago, they required meters for everybody. Now everybody has to pay attention to groundwater augmentation. Long term sustainability is the goal." A few ranchers who refused to install meters received heavy fines, to make it clear that the new policies and regulations would be enforced.

Colorado's realistic approach to long term sustainability contrasts with the gloomy forecast for the Ogallala Aquifer. The difference in effectiveness may be because the San Luis Valley is more compact and confined then the Ogallala, which spans so many states and groundwater jurisdictions.

Paul was also working on conservation easements along the Rio Grande, which moves across the southern edge of the valley through the Alamosa National Wildlife Refuge. At the nearby Monte Vista National Wildlife Refuge, we saw Sandhill Cranes (*Grus canadensis*) that had just arrived on their annual fall migration. After landing, they pointed their beaks to the sky and pranced in place. With a success story to share from the San Luis Valley, we felt like celebrating too.

At Wolf Creek Pass, after the highway climbed out of San Luis Valley, we reached the Continental Divide, and entered the region where waters drain toward the Pacific Ocean, instead of the Atlantic. If the gravel road to Summitville Mine had not been snowed-in, we might have turned off, but the mine was 16 miles south of the paved highway at 11,644 feet above sea level (37°26'N). Since December 1992, when the Canadian mining Company, Galactic Resources filed for bankruptcy and "walked away" from Summitville Mine, it has been an Environmental Protection Agency Superfund cleanup site. Like most modern gold mines in the West, they processed crushed ore in a "heap-leach" operation, piling it on top of a vinyl liner to be washed with cyanide to extract gold. At Summitville, the liner under a 160-million-gallon pond full of cyanide solution developed leaks. A dozen spills, totaling over 86,000 gallons, entered Cropsy Creek, a tributary of the Alamosa River, which eventually enters the Rio Grande. Fish died. Everything died. Faced with a $20 million cleanup job (the costs have since grown), and lenient national mining laws, Galactic left it to the nation's taxpayers to pay to clean up the cyanide pond and a major acid drainage problem.

This sad event actually contributed to the preservation of a historic mining town called Bodie, about 15 miles north of our home at Mono Lake. The same corporation, Galactic, had been preparing to mine the hill above the State Historic Park, where structures date back to the 1870s. A campaign led by the California State Park Rangers Association was underway, in 1992, to save the historic ghost town from the impacts of a modern industrial mining operation. When the mining company declared bankruptcy, the State Park System was able to purchase the hillside and those who cared about Bodie breathed a sigh of relief.

Our winding route passed through Durango, then swung north up a scenic valley full of aspen trees and back into the San Juan Mountains, through Silverton and Ouray, around another mountain range and up a box canyon to Telluride, a ski town on the 38th parallel that was a mining town in the late 19th century.

Victorian-style houses and a historic downtown district contrast with modern condominiums, while San Miguel Creek cascades along the edge of town. We planned to look into water issues left over from the mining era, then learned that a "350 Global Climate Change" event was scheduled for the day we arrived. That day, in over 5,200 locales in 181 countries (including every single nation and state touched by the 38th parallel), people gathered to highlight the goal to reduce carbon dioxide levels in the atmosphere from the current elevated level, about 390 parts per million, back down to 350 ppm, a level considered a safe upper limit to avoid runaway climate change. In Telluride, the New Community Coalition was forming the numerals "350" with bicycles, solar vehicles and hybrid cars.

The organizers needed another car to complete the figure, so our hybrid ended up front and center in the picture, forming the top of the number "5," in front of the court house podium. It was nice to briefly become part of that community and share a role in that worldwide day of climate change awareness.

UPDATE 2020

Colorado is planning for a future with less water and increasing "aridification" driven by climate change. 2017 and 2018 were drought years, somewhat relieved by a wet winter in 2019.

Colorado River Clean-up and Groundwater for Las Vegas

Entering Utah, we came downhill to the Colorado River at Moab, Utah, famous for mountain biking on slickrock trails and as a place to start river trips (38°34'N). A few years earlier we had canoed for a week on the Green River, through Canyonlands National Park to the confluence of the Green and Colorado Rivers, which is yet another landmark found by the 38th parallel. This time, we were in Moab to learn about the cleanup project underway to haul uranium mill tailings out of the Colorado River watershed.

Between 1956 and 1984, dozens of uranium mines operated 10 to 20 miles south of Moab, in outcrops of the Morrison and Chinle sedimentary layers (clustered very close to the 38th parallel). They sent their ore to a processing facility at Moab, because the Colorado River was the only water source in the region providing the consistent supply needed for milling. For 28 years, a tailing pile with radioactive uranium waste and high ammonia concentrations grew to cover 130 acres on the site; the edge of the pile was 700 feet from the west bank of the river, except during high runoff episodes, when the river sometimes touched that "toe."

"The pile actually started as a slurry pond and over time filled up with more and more sediment. It's kind of like a jelly donut," Elizabeth (Liz) Glowiak told us, "where all the fine materials traveled in toward the middle of the pond, and the sand on the outside. We call it the "slime" in the middle." Liz and her co-worker, Ken Pill are geohydrologists who, wearing bright yellow reflective vests, took us on a tour of the Superfund cleanup site. "The tailings are wet," Ken said, completing Liz's thought, "because they used up to 1600 gallons per minute of water diverted off the Colorado River to process the ore." Ken and Liz worked for a technical assistance team contracted to the Department of Energy, and smoothly took turns providing facts as we toured the Moab Uranium Mill Tailings Remedial Action Project.

From a viewpoint above the tailings, we watched dozens of trucks and loaders at work. Contaminated ground was being loaded onto trucks, which were thoroughly washed before transferring their loads to uncontaminated trucks (which never enter the pit); those loads then were delivered to train cars. There are special mechanics who work only on the dirty trucks inside. Twice a day, a loaded train hauled the contaminated material 30 miles north, for reburial outside the Colorado River watershed.

"Was the tailings pile lined when they were operating?" we asked.

"No, and that was typical practice in those days," Ken answered. "To give the Atlas Mineral Company a little bit of credit, a lot of other mining sites and milling sites were discharging directly into the river."

Millions of people who live downstream in Arizona, southern Nevada, and southern California rely on Colorado River water. If flooding washed over the tailings, or the plume moved underground into the river, the consequences would be devastating. Discussions about how to solve the problem had gone on ever since the mill company went bankrupt to avoid paying for the cleanup (the same tactic used by the mining company at Summitville, Colorado). "We took control of the site in 2001," Ken said. There was only enough money to consider capping the pile in place or to perhaps do some remediation. In 2010, once federal economic stimulus funds became available, the aggressive cleanup project began in earnest to provide a permanent solution. It will take 10 years of constant work to move 16 million tons, a literal mountain of tailings.

The mill used anhydrous ammonia to process the uranium ore. "Tons of ammonia," Ken said. It became the main contaminant in the groundwater and a concern for

Uranium mill tailings are being hauled away from the west bank of the Colorado River at Moab, Utah, to protect the water resource from radioactivity and ammonia pollution

fish habitat along the Colorado, because fish take up ammonia in their gills.

It was counter-intuitive to learn that low water flows in the river actually increase the likelihood of contaminated groundwater moving toward the river. High water episodes seep into the pile and create a counter-tension against the groundwater while diluting contaminates.

Ken and Liz monitor test wells across the edge of the site closest to the river. Besides uranium and ammonia, they sample for copper, selenium, manganese, and total dissolved solids in the brine, pumped from different well depths, and analyze their samples with equipment at the site. Wells also provide a way to proactively skim contaminates when concentrations are high. A high-pressure water stream shooting straight up in the air was an "air stripper," to convert ammonia to nitrogen gas. That required high temperatures and low humidity, which was exactly what the local desert climate provided.

At the far edge of the property, we walked through a riparian habitat that gave refuge to native Razorback Suckers and Tui Chub fish. Birds were singing within a strip of willows that could become part of a restored natural site at the gateway to Moab, after the tailings are gone.

"Will you be here for the entire ten years?" we asked Ken.

"I'd like to think so," he said. "I was down here when they first took over the site in 2002 and when they put in the first set of extraction wells. Now we're at the point where we're starting to do more remediation. So, I'd like to follow through on the whole thing."

In the middle of Moab, at the "Up the Creek" tents-only campground, we met the campground owner, Kimberly Schappert, who came to Moab over 20 years ago. She watched Moab change from a small mining town with a Uranium Cafe and Atomic Grill into a mecca for outdoor recreation, and became involved in county government, working for years to get the huge uranium tailing pile cleaned up. "That entailed meeting with the DOE [Department of Energy] people and the state DOE as stakeholders," she said, "and we would meet with the Metropolitan Water District of Southern California. We flew people out and took them on tours. Eighteen million Southern Californians and a bunch of Arizonians are what gave us the clout."

Money became the biggest stumbling block. "The original company, Atlas, had only a $6 million reclamation bond," she said. "Yet the cleanup project is up to $350 million now, and will be half a billion dollars by the time it is done. Atlas went bankrupt, so we taxpayers are footing the cleanup bill."

Kim had lots of idea for what could be done on the land once the cleanup is finished. "This has been so ugly, but it should become the great gateway to the community." She envisions a multi-use area, with "green" office space, parking for Arches National Park, and trailheads for mountain biking and walking. "This is a recreation area for people from all over the world, so whenever I apply for a grant, I keep in mind that we're not just serving the people of this county, but 2.5 million visitors a year from all over. This is their impression of the United States, so I feel like it's tax money well spent"

"What do you think it takes to see a problem like that and fix it? Is it key people in the community?" Janet asked.

"I think it is. People that have an issue, a cause, and are willing to fight for it. All the great things going on around here are not because of the government. It takes people..."

"People like you," Janet said.

"I have to admit, I work my butt off for this community, but other people have done that also, for other causes and issues. That's what gets it done. I'm sure that's what gets it done everywhere."

"You've inspired me to go home and try to make some things happen," Janet said.

Water and gravity have sculpted colorful sedimentary rock formations across southern Utah in some of the nation's most spectacular national parks. Our route included Arches National Park, Canyonlands National Park, Lake Powell National Recreation Area (behind Glen Canyon dam), Grand Staircase-Escalante National Monument, Capitol Reef National Park, and Bryce Canyon National Park.

On a bitterly cold afternoon, we arrived in the old mining town of Pioche, Nevada (37°56'N). In 1872, Pioche was the major town of southeastern Nevada. For two years, its population hovered around 10,000. The boom and bust pattern of western mining played out quickly here, as the principal silver mines closed in 1876. The town claims to have been even rougher than Bodie, the notorious mining town north of Mono Lake, because of frequent gunfights. Today it is the county seat of Lincoln County.

At the Pioche Library, we met Glennon Zelch, who was very involved in an on-going battle over Las Vegas's plans to export groundwater from eastern Nevada valleys. There are many parallels in this current water grab to the Los Angeles Aqueduct history that affected Mono Lake and the Owens Valley – family ranches being bought up for their water rights, local people overwhelmed by big city movers-and-shakers, and water moved hundreds of miles to serve development and growth in a distant, sprawling metropolis.

Glennon is a retired civil engineer who had worked with water utilities. When he moved to Pioche from Louisville, Kentucky, he was not aware of Las Vegas's pipeline plans. "I learned it was in the works when I stopped at the building department to see about a permit for my house," Glennon said, "and the lady in there asked me what I did, and I said I was a retired civil engineer. And she said, 'Oh goody, you're the only one in the whole county.'"

Glennon was asked to represent Lincoln County on a cooperative development committee that included a man from neighboring White Pine County, and 36 people from Clark County, which includes Las Vegas. "So, we were outnumbered 18-to 1 to start with," he said. "The whole purpose was to figure out a palatable way for the Southern Nevada Water Authority to put in this pipeline without stirring up too much public opposition. I went because I figured I could learn what was going on."

Las Vegas water planners have said they must have a minimum of 100,000 acre-feet of water out of White Pine County to make the pipeline project viable. One acre-foot is about enough water to cover a football field one foot deep and generally is sufficient as the annual supply for one household – or two if used conservatively. "They've bought about 50,000 of that," Glennon said, "and want to file on public land in Spring Valley and Snake Valley to get the remainder." Snake Valley could be the big stumbling block for the project. "If they pump from wells on the Nevada side, they'll pull water out of Utah. If Utah says "no" there's going to be hell to pay." If enough water did not materialize for Las Vegas in Snake Valley, the pipeline project would not be economical.

Though a tentative agreement had been reached between the states to share groundwater equally in the valley, opponents in Utah remained concerned that depleting the groundwater table will kill vegetation and lead to unhealthy dust storms blown all the way to Salt Lake City. Severe particulate air pollution developed in the Owens Valley and around Mono Lake after the Los Angeles Aqueduct was built, violating federal air quality standards and leading to expensive remediation requirements for Los Angeles.

"What do you see as the future for this town and this part of the county?" we asked Glennon.

"How much water you have dictates how big you can grow," he answered. "Los Angeles could not have gotten anywhere near as big as it is, if they hadn't stolen the water from elsewhere. Las Vegas is on the same track." We had heard that Las Vegas was promising that the northern valleys would have a viable economy, that there would not be 'another Owens Valley' episode because the city would share sufficient water.

"Well, we don't have a viable economy now," Glennon responded, "and if we lose our water, we can never develop one." The economy of Pioche, historically a mining town, is today based on ranching and government services, including the county courthouse, schools, road departments and a prison. "Without water, we're going to be welfare people. I told the people in White Pine County that I've met, 'File on every drop of water that you've got in your county, because if you don't, they'll get it sooner or later. Eventually they're going to extend the pipeline even farther north." Something similar had happened at Mono Lake, when Los Angeles decided to reach north of the Owens Valley. "That's the logical course of events," Glennon said. "They're not going to say, 'We've taken all the water we're going to take.' They'll just keep wanting more."

He finished with a provocative thought: "To be realistic, you should develop a community up to the limit of its resources. Then go develop a new community somewhere else. You can't just keep glomming all these people into one spot and trying to siphon water in from somewhere else to support them."

On the day we were in Pioche, a district judge overturned a 2008 State of Nevada ruling that had granted the Southern Nevada Water Authority permission to take groundwater from three valleys west of Pioche in central Lincoln County. Though an economic recession and a major housing slump in Las Vegas was dampening the thirst of that city, but there will be more chapters to this story in the years ahead.

Visiting Pioche was a particularly meaningful way to return home, after the around-the-world trip that had led us to so many water issues. The new aqueduct proposed by Las Vegas, that strangest of cities in the desert, has close parallels to our local Eastern Sierra history. The story brought us full circle, and now, after circling the globe, we were only a few hours away from Mono Lake.

As we crossed the state of Nevada, wild horses grazed near the highway. We watched for aliens along the "Extraterrestrial Highway," but eagerly pushed on. Wind was making waves on Mono Lake when we reached home, piling foam at the edge of the salty, alkaline water and whirling white flecks off into the sagebrush.

UPDATE 2020

The Moab Uranium Mill Tailings Remedial Action Project doubled its weekly mill tailings train shipments from two to four in February 2019. The shipping activities that began in 2009 are more than halfway to completing the goal to relocate 16 million tons of tailings.

In 2020, a Drought Contingency Plan took affect for the over-drafted Colorado River, negotiated by the river watershed states. Winter runoff flows had been low for most of the prior 20 years, with historically low reservoir levels in Powell and Mead. This new plan is a stop-gap until new operating guidelines are agreed upon in 2026.

Part IV: California: A "Water Line" to the Pacific

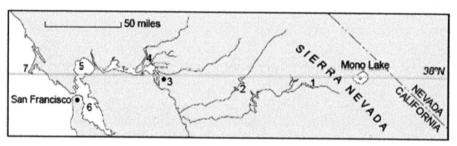

1. Hetch Hetchy, Tuolumne River   2. New Melones Reservoir, Stanislaus River   3. Stockton   4. Sacramento-San Joaquin Delta   5. San Pablo Bay   6. San Francisco Bay   7. Point Reyes

"The state of your inner nature is directly parallel to the outer nature where you spend time. That's why this is so much bigger than just monitoring birds of prey."

Allen Fish, Golden Gate Raptor Observatory

Mono Lake to the Sierra Crest

We were camped at 10,430 feet above sea level, just east of the Sierra crest, looking at small glaciers clinging to the north slopes of Mt. Conness and North Peak. Wispy clouds turned pink and there was the beginning of alpenglow overhead. It was our first evening on a 17-day trek across California, continuing along the 38th parallel from Mono Lake to the Pacific Ocean, exploring the mountains-to-the-sea watershed that sends water to San Francisco Bay and, finally, out to sea.

At the slow pace dictated by travel on foot, bicycle, and boat, we traced a "water line" where battles have been fought over dams, aqueducts, and wetlands, and where critical water issues still are being played out. The California list includes Mono Lake, the snowpack at the Sierra Nevada crest, Hetch Hetchy, New Melones Reservoir, the Sacramento-San Joaquin Delta, San Francisco Bay, Point Reyes, and the Farallones Islands. In each location, the water issues of our home state of California had parallels and contrasts with discoveries we had made at lakes, rivers, marshes, estuaries, and islands along the 38th parallel, around the world.

Our starting point at Mono Lake, east of the mountain range, seems hydrologically separate from California's west slope watershed, until you consider the intricate water system that spans the state. Most Californians are tied together by man-made plumbing, ensuring that stresses on the water supply system in one place are today felt throughout the state.

Those connections began when the City of Los Angeles, in 1905, asked its citizens to approve a bond for construction of the Los Angeles Aqueduct from the Owens Valley, hundreds of miles to the north. Tapping the Owens River for water allowed the city to grow far beyond the limits of its own local water sources. Thirty years later, the city reached even farther north, adding streams from the Mono Lake basin to its distant water supply.

Much more population growth came to the southern California region after the Colorado River aqueduct began delivering even greater quantities of water to the region in 1941, and after 1960, when voters approved construction bonds for the dams and aqueducts of the California State Water Project. The California Aqueduct carries runoff water 444 miles from dams in the northern Sierra to southern California cities as far away as San Diego. Farther north, in 1913, the city of San Francisco received Congressional approval to build a dam on the Tuolumne River in Hetch Hetchy valley, inside Yosemite National Park. And, for the East Bay, an aqueduct spanned the middle of the state to tap mountain runoff from the Mokulumne River. Meanwhile, beginning in the Great Depression years, the federal government built the Central Valley Project, ultimately totaling 20 dams on five of California's major rivers, plus aqueducts to transport northern California water to farms in the southern San Joaquin valley. Today, with over 1,400 dams and a spider web of canals, water choices made by any section of the state may have repercussions hundreds of miles away. Key elements of California's 21st century water web lie along the 38th parallel, including Mono Lake.

At a sendoff party at the lake, we floated in the salty water and told a group of friends about our hopes for the trek across the state. We had worked as rangers at the Mono Lake Tufa State Reserve for more than two decades, in partnership with the Mono Lake Committee and U.S. Forest Service. It took the Committee's leaders and lawyers, backed by 20,000 members, 16 years to build public concern and win battles in court (partnered with the National Audubon Society and CalTrout) that ultimately protected the lake and its tributary streams from the effects of stream diversions into the Los Angeles Aqueduct. A key victory came in 1994, when the city's water licenses were amended to reduce diversions from Mono Lake and gradually raise its volume back to where the unique saline

Limestone tufa towers grow beneath salty, alkaline Mono Lake, where more than a million birds feed and nest east of the Sierra Nevada

ecosystem remains healthy, teeming with brine shrimp, alkali flies, and birds, and to end dust storms blowing off the exposed lakebed that violated legal air quality standards. Ubiquitous "Save Mono Lake" messages on bumper stickers were replaced by "Long Live Mono Lake."

The key court decision relied on the Public Trust Doctrine, which recognizes the government's legal responsibility to protect resources that have environmental and aesthetic values – resources that are held "in trust" for the public. Applying that doctrine to Mono Lake altered water rights that had been held for a half century by the city of Los Angeles, and set a legal precedent that rattled water agencies facing similar issues elsewhere. Yet, the city was not devastated; instead, significant water conservation and recycling measures were put in place to save more water in Los Angeles than they had been diverting from the lake.

We had witnessed an environmental victory that contrasts vividly with the Aral Sea devastation. In Turkmenistan, no "Aral Sea Committee" formed to advocate for protecting the resource values of that massive inland sea before river diversions, to serve the distant city of Ashgabat and regional farms, destroyed the ecosystem and subjected the region's population to harmful dust storms. And no such legal too is available to the people of eastern Turkey, to protect Hasankeyf on the Tigris River. Nor has there been a successful legal or public opinion campaign, yet, to protect the diminishing Ogallala groundwater resource, perhaps (as we heard in western Kansas) because that problem is hidden from sight.

We asked Geoff McQuilkin, the Executive Director of the Mono Lake Committee, about the lake's connections with water environments west of the Sierra. Reducing diversions from Mono Basin streams required Los Angeles to conserve water, which, Geoff noted, "affects how much is drawn out of the Bay-Delta for Southern California and in turn affects users right there along your route. It's all linked together. We want to see good things happen in the Sacramento-San Joaquin Delta, where you are headed, because that means there will be less reason to come back and seek more water from Mono Lake." Geoff even saw connections tracing back through the millennia. "Millions of years ago this was part of the San Joaquin River drainage out to the ocean, before Mammoth Mountain and the Sierra crest rose so high." The Sierra Nevada is still being forced upward by colliding tectonic plates. Near the end of our California trip we would cross the San Andreas fault and leave the North American plate behind as we walked onto the Pacific plate at Point Reyes. Hydrology is a function of topography, and topography would be on our minds in the coming days, as we faced the climb over the Sierra Nevada escarpment.

We took a deep breath, then walked out the door of our house a mile north of Mono Lake (38°02'N; 119°08'W), heading west up Lundy Canyon, where Mill Creek gathers snowmelt from the surrounding peaks before flowing below our house into the lake. Lundy Canyon is full of aspen groves, dramatic wildflowers, and waterfalls. The canyon steadily angled southwest and, at the top, we stood exactly on the 38° line.

Connie Millar appeared out of the wilderness, moving at a brisk pace as the afternoon sun dropped toward the Sierra crest. Short and wiry, with a head of curly red hair, our neighbor is a U.S. Forest Service paleoecologist who explores

North Peak glacier, shrinking due to a warming climate

connections between mountain vegetation and changing climate patterns. She told us that photographs taken 80 years ago documented how the nearby glaciers had shrunk by at least 80 percent during the last century. "North Peak Glacier may be gone in a decade," Connie told us, "and all of the glaciers in the central Sierra in our lifetimes."

As we had seen in western China, some of the clearest signs of climate warming are showing up at high elevations. Scientists like Connie are trying to understand what global changes mean for local areas, because effects vary. "In parts of the eastern U.S., temperatures are actually cooling," she told us, "while the West is warming much faster than the global average." This kind of complexity confuses some people who expect to correlate the big picture with what they see happening outside their homes.

A rising temperature trend has been recorded in California through the last 120 years, as the industrial age tapped into the carbon trapped in fossil fuels, but the slope of that curve steepened over the last 30 years (coinciding with a population doubling that pushed human numbers beyond 7 billion). And "in the last 10 years everything has shot up," said Connie. It becomes evident that human release of greenhouse gases is driving the changes, after analyzing all the variables that naturally alter climate, such as fluctuations in solar radiation, variations in the Earth's rotation, and volcanic eruptions. The observed trend is only matched when computer models include the elevated greenhouse gases released by human burning of fossil fuels.

Forecasts give two-to-one odds that much of California will be drier by the end of the century – a 20 percent decline in precipitation – with more water falling as rain, less as snow. A trend of earlier snowmelt runoff, shorter spring seasons, and longer summers is already apparent in the Sierra Nevada. These trends are similar to those projected for the western Mediterranean nations, that we had heard about while in Portugal, and in both regions will complicate the timing of water storage and release by reservoir managers.

We were sitting near the treeline. Scattered clumps of trees were primarily whitebark pine and some limber pine. Connie explained how "drier" added to "warmer" was killing trees. Higher "evaporative stress," as Connie put it, was "pushing trees over thresholds they had not experienced in the last several hundred years, with hillsides going in a flash of mortality. But even where there is a lot of mortality, it's not total." Slower-growing pines were being taken out, producing strong selection for trees that may do better under the altered climate.

We all wondered what might happen to Mono Lake. It would be tragic if the lake's ecosystem collapsed due to climate change. The inland sea survived prehistoric droughts, but the buffer taken away by 50 years of stream diversions has not yet been restored.

Connie's concerns were broader, stretching from Mono Lake to even the smallest riparian corridors. She has a special interest in "rock glaciers," where ice lies embedded beneath insulating rock, something we could see at the lower edge of North Peak Glacier, just south of us hugging the mountain crest. The terminal moraine there was seeping water. Rock glaciers, with insulating coats of rubble, should help wetlands stay lush and alive for many years during a warmer climate regime. "They won't fill Mono Lake up,"Connie explained, "but will serve these local wetlands and provide persistent streams where other canyons will just dry up. It's encouraging, not for the statewide water supply, but locally for the birds, and wildflowers and pika." Pika, (*Ochotona princeps*), are small rodents that gather grass each summer to dry in "hay" piles for fodder during the long alpine winters.

California's official climate strategy is to prioritize efforts toward the most sensitive resources. "People are talking about triage these days," Connie said. "There will be things we may have to let go. On the west slope of the Sierra, almost all of the resource plans have a priority to re-introduce salmon. But by mid-century, waters may become too warm to support natural salmon runs, so you may not want to put the effort there if you are just going to lose in the long run."

It was a grim picture to consider as we sat in one of the world's most heavenly settings. There may still be time to modify this scenario, if political will, here and abroad, addresses the human pressures driving global changes in the atmosphere and oceans. International agreements seem ineffective, so far, despite considerable renewable energy and low-carbon emissions transportation measures we had seen in parts of Asia, Europe, and in our own country.

On a beautiful bluebird morning, we followed a trail up to Secret Lake. We remembered seeing mayflies hatching there in August, coming up like magic from the depths and appearing on the water surface, then popping their wings out to dry. The flies meant lots of protein for finches that snapped them off the water surface. At the Sierra crest we paused to gaze westward across the wilderness lands in Yosemite National Park that we would travel in the coming days.

UPDATE 2020

California experienced a severe drought from December 2011 to March 2017. Smaller winter snow packs reduced runoff to Mono Lake and the recovery there stalled, with more than 10 feet of rise still to accomplish to meet long-term management objectives. What might have taken 20 years, under past "average" precipitation patterns, may now extend over another decade, depending on how the climate crisis plays out.

Grand Canyon of the Tuolumne to Hetch Hetchy

Loaded with heavy backpacks, we slipped and slid down a rough trail to the McCabe Lakes, where we found fish nets stretched out across the water. The National Park was two years into a five-year project to clean fish out and improve conditions for native Yellow-legged Frogs. We saw no fish, but also no polliwogs or adult frogs, yet. It is easy to introduce fish into a lake, but very hard to get them out.

Sierra Nevada frogs are beleaguered not only by trout predation, but also by a fungus that is infecting amphibians all over the world and has spread across much of our mountain wilderness, perhaps carried by flying insects. Some individual frogs will hopefully have resistance, but add in pesticides, which interfere with reproduction, blown on the west winds from Central Valley farms, and the amphibians' prospects look bleak.

In the following days we descended through the breathtaking Grand Canyon of the Tuolumne river, dropping past California, Le Conte, and Waterwheel Falls, gravity pulling the water, and us, down and down. We left the lodgepole forest behind and walked beneath black oaks, and Jeffrey and sugar pines. The canyon featured a series of cascades and pools, with massive granite walls framing the views and channeling our direction of travel through a landscape that resembled Yosemite Valley (without tourists). We saw almost no other people.

Above 6,000 feet, it felt like everything was in a hurry to set seeds before winter, including mountain ash, with its clusters of bright red berries. Where we first encountered oak trees, acorns had been falling under canopies of yellow-brown leaves, but farther down canyon the trees had barely begun to think about autumn.

A long cascade into an emerald pool got us to stop for a dip. We had it all to ourselves. In another cascade of pools, a Dipper (Water Ouzel) was feeding under water. It was a great place for both human and avian dippers.

Cascades fall toward the Grand Canyon of the Tuolumne River in Yosemite National Park

The canyon opened wider at Pate Valley where the elevation was only 4,350 feet (37°56'N). The air felt noticeably thicker and the trees were mostly oaks and incense cedar. Downstream from there the canyon curves south and then back west toward Hetch Hetchy Valley. No trails follow the upper shores of Hetch Hetchy Reservoir, so we instead faced a strenuous two-day detour. During Congressional debates over construction of a dam inside Yosemite National Park to serve as a City of San Francisco

Above Hetch Hetchy reservoir in the Yosemite National Park wilderness

water source, Secretary of the Interior James R. Garfield had said, "the prime change will be that, instead of a beautiful but somewhat unusable 'meadow' floor, the valley will be a lake of rare beauty." (Egleston, 1909, 7). Plans were made for lodges and roads and trails, but once the dam was in place the no-swimming or lakeshore access policy was instituted so that city water would not have to be filtered. Or, perhaps it has simply been in San Francisco's best interest to keep awareness low about changes their dam brought to Yosemite's twin valley.

We left summer behind in Pate Valley on the 3,500-foot climb to Harden Lake, and returned abruptly to early autumn at 7,600 feet (37°53N). There were day-hikers there and though we had 12 backcountry miles to do the next day, running into people made it clear we had finished the most remote, wild part of our crossing.

Navigating switchbacks on the trail down toward Hetch Hetchy, we glimpsed what struck us, at first, as a big slab of gray granite. Then our wilderness-focused brains adjusted and we realized it was the paved road to O'Shaughnessy Dam.

In 1913, Congress authorized the dam that flooded Hetch Hetchy Valley, which is about two-thirds the size of world-famous Yosemite Valley, where the Tuolumne River meandered through flowery meadows and dramatic waterfalls plummeted off the cliffs of towering granite domes. John Muir, leading the unsuccessful fight against approval of the dam inside Yosemite National Park, wrote the famous lines: "Dam Hetch Hetchy! As well dam for water tanks the people's cathedrals and churches, for no holier temple has ever been consecrated by the heart of man." (Muir, 1912, 181).

In the backpacker's camp not far from the dam, Spreck Rosekrans found us. He is chairman of the board of Restore Hetch Hetchy and also handles water issues for the Environmental Defense Fund (EDF). He became our new best friend when we saw he was carrying fresh food and wine.

Tall and fit, Spreck exudes contagious energy. Restore Hetch Hetchy hopes to remove the reservoir from the valley, an idea that seems audacious to many people, particularly the San Francisco Public Utilities Commission, given the water supply challenges facing California. Yet feasibility studies, including two by the National Park Service and U.S. Bureau of Reclamation in 1988 and the California Department of Water Resources in 2006, concluded that existing storage downstream can feasibly replace the water and most of the hydroelectric power from Hetch Hetchy. But in the State of California report, Spreck explained, "they reassessed cost numbers prepared by the Environmental Defense Fund [that estimated a $1.5 billion cost] and said we needed 30 percent more for engineering. They called for replacing the water by four times over! Their approach would keep a water bank in Don Pedro Reservoir, keep Cherry and Eleanor reservoirs, then add another dam (in the coast range), plus groundwater storage, plus conservation, plus purchasing water from willing sellers. They called for *all* those things and concluded it would cost between $3 billion and $10 billion!

"Well, we said you could do one or the other or a mix. We want an academic, independent body to oversee these alternatives. The state said this study was as far as they were going to go and kicked any future action over to the federal government."

Restoration of Hetch Hetchy was given considerable federal support during President Reagan's administrations. "Don Hodel, Secretary of Interior under Reagan, made the suggestion in 1987 that restoration could occur with no harm to San Francisco," Spreck said. "Don is not someone you'd think of as a 'tree hugger,' but he understands the natural heritage here."

The Restore Hetch Hetchy organization formed in 1999 as a split-off group from the Sierra Club. It would have pleased John Muir to see this century-old battle re-engaged. Muir's last major conservation battle had been over the dam. He died of pneumonia in 1914, the year after the Congressional authorization, but history has concluded that Muir died of a broken heart over the loss.

"The damming of Hetch Hetchy was the event that turned the Sierra Club from an outing club to a political organization," Spreck said. "Two years after Congress authorized this dam, they passed the National Parks Act, basically ensuring that we're going to preserve parks and not do anything like building the Hetch Hetchy dam, ever again. The National Park System might not exist except for what happened here at Hetch Hetchy."

For us, knowing it is feasible to restore the sister valley to Yosemite without a loss of San Francisco's water supply, it seems like a moral imperative to correct the hundred-year-old mistake – a gift worth pursuing for our children and the world. Damming and flooding a natural treasure like Hetch Hetchy was a bad decision one hundred years ago, that is similar to the choice being made in Turkey, today, to flood the

200

archaeological and historical treasures of Hasankeyf. In both cases, short-sighted objectives of local dam-builders seem blind to long-term values that matter to the entire world. Correcting the Hetch Hetchy mistake may take many years but remains feasible; there is a short window of opportunity to avoid a similar mistake on the Tigris River in Turkey.

"One possible approach for our future effort," Spreck concluded, as we cleaned up after the meal, "is a clause in the California Constitution that says all water will be 'reasonably and beneficially used and methods of diversion shall be reasonable.' So," he asked, "is it reasonable to divert water into storage inside a National Park?"

In the morning, Spreck walked with us across O'Shaughnessy Dam and we looked at the exhibit panels. It seemed wrong that there were no photographs of the valley as it appeared before the dam. Spreck did us another major favor by driving Janet and our backpacks to Cherry Lake while David walked the 15 miles, carrying just a day pack. Cherry Lake was the one part of the San Francisco water system that Spreck had not ever visited. Janet found a campsite and put signs up for our son, Ryan, who was meeting us there with bicycles and supplies.

The trail that David followed (actually an overgrown construction road used to haul gravel to the dam site during construction) passed through oak and pine forest toward Eleanor. That "lake" was nearly full, which always helps a reservoir's looks. Along with O'Shaughnessy Dam and the aqueduct facilities delivering water to San Francisco, the entire Hetch Hetchy system took 20 years to complete after the 1913 Raker Act. Water from Eleanor Lake moves into Cherry Lake, which passes it down through hydroelectric generators and eventually into the Tuolumne River. Boating and the typical recreational uses in reservoirs *are* allowed in those two man-made lakes, unlike at Hetch Hetchy, feeding our suspicion that the access restrictions have political motives.

We reunited at Cherry Lake campground, at 5,000 feet (37°58'N). The mountains were behind us.

UPDATE 2020: Restore Hetch Hetchy lost in a lawsuit against San Francisco in 2015. Their efforts continue.

Friends of the River and New Melones Reservoir

On the autumn equinox, we bicycled down through the Stanislaus National Forest, aware that all of us on the 38°N latitude line, around the world, saw the noon sun that day at 52 degrees (90° - 38°) above the horizon. Our 40-mile ride to Sonora was an extremely hilly, forested, and scenic route. At one point we were pedaling so slowly uphill that a butterfly heading the same direction passed us by.

One of the canyons bottomed out at the Clavey River, an unusual water body because its direction of flow is north-to-south rather than the typical east-to-west orientation of Sierra Nevada streams. Though there was graffiti on the bridge and an official "no bungee jumping" sign, this tributary of the Tuolumne River is actually one of the most pristine rivers in the Sierra Nevada. It flows dam-free for its entire length down to the Tuolumne River, one of only three Sierra streams without a major diversion or dam. The unique Clavey Rainbow Trout lives in this state designated "Wild Trout Stream" and Forest Service "Critical Aquatic Reserve," yet, so far, the Clavey had not been given protection as a federal Wild and Scenic River.

In the foothill town of Sonora, at Kurt Stegen's house, we took our first showers in nine days, then Kurt drove us to the New Melones Reservoir visitor center. He is an arborist who had been a volunteer with Friends of the River for 35 years and personally experienced the contentious history of the Stanislaus River. Seeming, at first, quiet and serious, we soon realized Kurt's passionate resolve.

New Melones, completed in 1978, became the final dam built by the Army Corps of Engineers for the federal Central

Valley Project. When first proposed, back in 1944, it was simply intended to control floods, but Congress modified its authorization in 1962 by adding irrigation, hydropower, fish and wildlife enhancement, recreation, and water quality to the list of projected benefits. Contention about New Melones was not about whether to build the dam, but how high it should be and how much of the river canyon it would flood. In 1974, Friends of the River sponsored a state ballot initiative, Proposition 17, to prevent the reservoir from inundating nine miles of a popular river rafting stretch at the upper end of the canyon.

"I think the most incredible thing about Friends of the River was the grassroots campaign they put together," Kurt said. "There was no internet, of course. I don't really understand how we communicated, but it was a well-organized campaign." In the 1970s, newsletters were the medium of choice, prepared on typewriters, using lots of correction fluid and making headlines with rub-on transfer letters, then photocopied before sending via "snail-mail" (a term that could not exist until the internet came along).

Another group, Friends of New Melones, formed to defeat the ballot measure and confused voters with billboards that read: *Stop "Wild River" Hoax! Stop Pollution of the River!* That a new reservoir could release water in the summer to dilute downstream pollution justified the anti-pollution message. The "wild river hoax" claim was made because upstream dams controlled the river's natural flow. (Of course, many wild rivers share that status, including the Colorado River through the Grand Canyon.) The Los Angeles *Times* editorialized that "the billboards seem an absolute betrayal of the truth to us. We too, oppose Proposition 17. But to call it a 'wild-river hoax' and to suggest that the proposition would result in pollution is a resort to tactics that have no place in responsible democratic campaigning." (Los Angeles Times, 1974, 6). Yet the tactic proved effective; Proposition 17 was defeated.

An exhibit at the visitor center summarized some of that history, which Kurt and so many others lived first-hand. From there we drove to the viewpoint overlooking the dam, a massive earthen plug 625 feet high, spanning 1,560 feet across the canyon. Kurt regularly takes school groups to the

New Melones reservoir is on the Stanislaus River
in the Sierra Nevada foothills

viewpoint, but a gate usually keeps the road closed and we were the only people to walk in that day.

In 1979, as the reservoir began to fill, Mark Dubois, then director of Friends of the River, chained himself to a boulder in a secret location that would be flooded. Several others joined that protest, forcing water to be released from the dam to avoid drowning the protesters.

Though the ballot initiative had been defeated, the Friends of the River campaign did produce considerable political support in Sacramento. In May 1979, Governor Jerry Brown vetoed a bill that would have allowed New Melones reservoir to be filled to capacity. In 1980, the California State Water Resources Control Board set a low limit for the reservoir level, but heavy runoff after a record winter in 1982 trumped political will and filled the reservoir to its spilling point. In 1983, the Board lifted its restrictions; a full reservoir was a *fait accompli*.

That record winter snowpack of 1982-'83, coincidentally, followed the summer we arrived to begin work at Mono Lake.

We had expected to watch the lake continue its long decline, but following the exceptional winter it rose six feet, postponing the ultimate crisis point for the lake ecosystem. The dry stream channels feeding Mono Lake again ran with water that could not be handled by the Los Angeles aqueduct system, but instead of defeating environmental goals, as happened at the New Melones Dam, the re-watered channels triggered efforts to keep that water flowing and fish alive in Mono Lake's tributary streams.

As for the long list of projected benefits from New Melones, it has not met expectations. A history of the project prepared in 1994 for the Bureau of Reclamation concluded that this particular dam could be considered "a case study of all that can go wrong with a project." (Simonds, 1994). There has not been sufficient water to meet all of the obligations used to justify the project.

In Korea, China, Turkey, Spain, and Portugal we had learned about the other side of the balance sheet, the long list of problems that dams sometimes create along with their benefits. Dams destroy natural processes of rivers. They trap sediments that would build beaches where riparian vegetation and wildlife thrive, and block salmon and steelhead trying to reach ancestral spawning grounds. Flooding's beneficial aspects – scouring sediments, replenishing spawning gravels, and fertilizing floodplains, are lost. And so, the Quercus group in Portugal called the Alqueva dam a "white elephant."

Though Quercus and the Friends of the River were defeated in their respective campaigns, both organizations moved on to successes. Friends of the River campaigns helped protect many of California's major rivers, including the American River where construction of the Auburn Dam was stopped due to seismic risks. In the mid-1980s, they founded the California Water Caucus, a statewide coalition to protect the Sacramento-San Joaquin Delta and promote environmentally sound water policy.

Kurt "stumbled onto the Stanislaus River," he said, "when I was 18 years old, through a job. In 1973, I didn't realize how involved with it I would become. I don't usually get caught up in things, but it had a special appeal to me." "Stumbling" into a commitment that ends up shaping a lifetime of concern is a pattern that would be recognizable to other activists we met, around the world.

Toward sunset, we stood at the upper end of the reservoir where the whitewater rafting stretch once began, and looked at tree snags poking above the water (the reservoir was only at 46 percent of capacity that autumn). For Kurt, that low level suggested an intriguing possibility: might the reservoir be managed to restore the whitewater rafting stretch, while still ensuring emergency flood storage capacity in high water episodes like those forecast to occur more frequently with global warming? Bringing back flowing water to the Stanislaus River may not be an impossible dream.

Leaving the foothill town of Sonora, 12 miles of bicycling through the Gold Rush country brought us to Angels Camp. Samuel Clemens spent time in that little town in the 1860s (after first visiting Mono Lake) and heard a story in a saloon that inspired him to write, "The Celebrated Jumping Frog of Calaveras County," his first story published as Mark Twain. Since 1928, Angels Camp has celebrated the annual Jumping Frog Jubilee each May. Unfortunately, the native Red-legged Frogs that inspired the event are nearly extinct in the nearby foothills, partly because of predation by Bullfrogs introduced from the East. Today's jumping competitors are all the much larger Bullfrogs.

From Angels Camp we turned west again, dropping out of the foothills through Salt Springs Valley, a "lost" valley off the usual travel routes. Scattered ranch houses, cattle and fields exist in a part of this crowded state seemingly bypassed by "progress." Just east of there, by contrast, retirement communities and golf courses sprawled across the lower foothills. We saw White Pelicans, Pied-billed Grebes, and assorted ducks in the reservoir at the lower end of the valley. As we approached Hwy 26, the road we were traveling was lined with gravel tailing piles left by miners who had seemingly chewed their way across the land, historic remnants similar to the present-day impacts on the Jade River in Hotan, China.

The hard, physical part of this trek was over once we reached Stockton, since we would embark by boat in the morning. Hiking and biking had given us a new, physical appreciation for the size of California.

The Sacramento-San Joaquin Delta

Perched on the dock with our gear, we watched a parade of boaters come and go: mom, pop, dog, then several lone fishermen and five head-shaved, tattooed young men crammed into a boat built for speed rather than fishing. John Knott's 33-foot Catalina sailboat emerged from among a group of college rowing-crew sculls. It was to be our "mother ship" for the next five days.

John, an experienced sailor in those waters, had been the Sierra State Parks District Superintendent, which included responsibility for Mono Lake. Now retired, he brought along David Martin as crew, as they sailed from San Francisco Bay to meet us. John's brother, Marty, sailed along with us in his smaller catamaran. It was sunny, hot, and amazingly clear. That afternoon we boated only a few miles in the Stockton deepwater shipping channel. Shortly after leaving the dock, we intersected one of the proposed routes for a controversial peripheral canal that could mean even less water flowing through natural Delta channels, by redirecting it around the periphery of the estuary to the southern aqueducts.

The Sacramento-San Joaquin Delta gathers the waters of California's two major river arteries into the largest estuary on the west coast of the United States, with 1,000 miles of channels that feed that water toward San Francisco Bay and the Pacific Ocean. About 75 percent of California's water demand from cities, suburbs, and farms originates south of this point, while about 75 percent of the precipitation that falls in the state occurs to the north. To create 21st century California, water must be shifted from north to south in aqueducts served by massive pumps at the south edge of the Delta. Populations of fish – salmon, smelt, and shad – dependent on natural Delta flows, have plummeted as water diversions altered the ecosystem, especially after exports increased from a yearly average of 4.6 million-acre-feet, to 6 million-acre-feet between 2000 and 2007. The damage to the estuary ecosystem and the competing "thirsts" of cities, commercial fisheries, and farms make the Delta not just a key transfer point, but a hub of the state's never-ending political and environmental battles over water.

Toward sunset that first night on the Delta, we turned out of the deepwater channel and anchored near Lost Isle. David Martin had a portable GPS unit; he paddled his kayak a few yards away, then called back that he was *exactly* on the 38th.

It was a beautiful evening in a natural setting. Swimming in the warm water felt fantastic, though we could not help speculating about the agricultural runoff chemicals that were in the mix. A U.S. Fish and Wildlife biologist has called the lower San Joaquin River "the lower colon of the valley's river system" because it carries so much pollution. (Carle, 2003, 145).

With the sky glowing orange behind Mount Diablo's silhouette, we also tried out the kayaks that traveled behind the sailboat on leashes. One side of the channel was green and lush and noisy with birds. The other side was a bare rock levee, devoid of life (part of 1,600 miles of levees lining the Delta channels). As night fell, from behind that levee we heard the insistent beeping of vehicles backing up. The next morning, we saw the source of the noise: trucks driving atop the levee, each pulling two trailers brimming with tomatoes harvested during the night.

Mount Diablo had been visible ever since Stockton, and would be with us all the way into San Francisco Bay. The peak is also a particularly important landmark for California geographers. Though it does not sit atop an intersection of exact latitude and longitude lines (37°51'N; 121°55'W), Diablo's 3,845-foot summit looms high above other coast range hills. On the clearest days, the Sierra Nevada and Mount Shasta are visible from there. That explains why, in 1851, Colonel Leander Ransom, a surveyor with the General Land Office chose Mount Diablo's peak as an "initial point."

He set a marker on the summit and then used it to survey prime meridian and base lines. Working outward from those lines, surveyors had divided California and Nevada public lands into townships (squares with six-mile sides). The perpendicular angularity of many of the Central Valley's roads and fence lines reflected those imaginary lines running across the landscape, originating from Mount Diablo.

Life on the river is a whole different world, with pace dictated by tide, wind, and marina. It was hard to believe that the little channel in front of us could take us to the Golden Gate and the open sea, but as we chugged out the gradually widening channel, the scale of the Delta revealed itself. We do not have many big rivers in California, but the Delta has a taste of the Mississippi or Ohio Rivers. We could feel the power of the water, full of life, giving life, moving life. It had a wonderful smell, like "sweet rain," Janet decided; a very different feel and odor than a Sierra stream, richer and dense with life.

On our first full day on the river, we saw a dark head, the thrashing body of a Sea Lion, then a flash of a huge silver fish. It was an exciting glimpse of a sight becoming rare in the Delta as salmon numbers have declined. Conditions were very different a half-century ago, when the rivers teemed with multiple salmon runs, passing far up into the mountain waters to spawn and add their bodies to the river food webs. California has four distinct runs of Chinook Salmon, named according to when they pass through the Delta on returning from the sea. The winter run species is endangered, the spring run is threatened, and the two fall runs are "species of concern," according the U.S. Fish & Wildlife Service. Millions of Chinook used to swim upriver each year; in the winter of 1883-'84, over 700,000 were caught for Delta canneries. Dams constructed in the 1940s, '50s, and '60s dealt a severe blow by blocking access to the upper reaches of Sierra Nevada rivers and streams; only 10 percent of suitable spawning habitat remains. Less water flowing out to sea carrying the chemical signatures of each salmon's birth water has, more recently, brought the species to the brink.

Midday, we paused near Frank's Tract for more exploring in the kayaks. Sixteen miles straight south from there, pumps serving the California Aqueduct sent water toward Southern

California cities, and others into the Delta-Mendota Canal for farms in the San Joaquin Valley. If we let the boat drift with the current, we might have been pulled toward the pumps instead of downstream, because their suction is powerful enough to reverse flows in Delta channels.

We had been given an update on issues affecting the Delta when we were at Hetch Hetchy, as Spreck Rosekrans had just attended a Bay Delta Conservation Plan (BDCP) meeting. The BDCP group, appointed by the Governor, sought agreement among water diverters, environmental interests, and Delta farmers and residents, a process Spreck characterized as "gnarly." Water agencies want the peripheral canal built around the Delta, or a canal beneath it, so they can continue moving freshwater if levees fail and saltwater reaches their pumps. Many of the 2,500 Delta residents worry that, with a canal in place, the other interests would no longer share the "common pool" that gives everyone an incentive to maintain levees. Major concerns remain that enough flow be guaranteed to the estuary ecosystem and that a peripheral canal not simply facilitate ever more diversions.

One canal proposal envisions massive tunnels beneath the Delta that might not be full except during the wettest, surplus-water years, but would be capable of shifting lots of water to south state reservoirs. That much capacity, however, was worrisome: California's development history makes it hard to trust loosely-worded assurances that sufficient water would run out to San Francisco Bay at all times.

We sailed past several restoration projects trying to bring back some of the original tules, seeming very small against the scale of acreage that has been cleared, plowed, drained, and controlled. For a short while that afternoon there was enough wind to put up the sails, though the sailors kept grumbling that the calm weather was *too* fine throughout our cruise. Our second night on the boat was at the marina at Brannon Island (38°07'N).

John Cain met us there the next morning. He was the director of restoration programs for the Natural Heritage Institute (in the 1990s, John had worked for the Mono Lake Committee). John's dark eyes lit with enthusiasm as he described his work. The Natural Heritage Institute protects and restores aquatic ecosystems and the services they provide to humans. Our time with John had to be early in the day, because he had an afternoon BDCP meeting in Sacramento as co-chairman of a habitat restoration committee.

While we toured in John's car, the boats shifted downstream. Our road ran beside the Sacramento River where massive wind turbines decorated the hills beyond the channel. "This channel of the Sacramento River is artificial," John said. "Ships used to get stuck in the Gold Rush days so they dredged here and dumped the deposits along the edge. Now, the deposits are the only soil that can be taken for rebuilding islands or strengthening levees around the perimeter of islands."

Farm soils on Delta islands subside once the peat soil, naturally underwater in the marshes, is exposed to air. Oxidation converts complex carbon compounds into carbon dioxide, causing one to two inches of lost soil depth each year. So, most Delta islands are today below sea level, protected by surrounding levees and serviced by groundwater pumps that constantly run to keep fields dry enough to grow crops. Electricity to run the pumps can be a farmer's biggest expense.

On Twitchell Island, 12 feet below sea level, native tules and cattails were growing on several acres of ponds. John reached into several inches of water and came up with a handful of dripping black mud. "The rate of tule muck accumulation is about two inches per year. We're reversing subsidence – actually building up this island. These subsiding islands had become a substantial source of atmospheric CO_2, so this converts a carbon source to a carbon sink. We're hoping to repeat this on a much larger scale."

We mulled over "carbon-capture farming" as we drove to the south end of the island to see an experimental "green levee" covered with lush growth. Engineers prefer that levees be bare so they can see signs of failure, but during a recent flood event, John said, "when people thought the standard levee was going to go, everyone took shelter behind this levee armored with vegetation."

Levee failure on an island in the north part of the Delta could send salt all the way to the aqueduct pumps at the south end. "When a big gulp of salt water suddenly moves in," John explained, "it can lead to a chronic problem. Not all of the salt on a breached island leaves when the tide goes out. That salt

John Cain held a handful of peat soil built by decaying tules
and cattails in the Sacramento-San Joaquin Delta

can be dissolved back into the water and gradually shifted by high tides farther and farther toward the aqueduct pumps." Salt water, of course, is useless to farmers or for drinking water, so intact levees are critically important.

We crossed Sherman Island, 11,000 acres at the western margin of the Delta that was theoretically protected by levees, but the walls were built on sand foundations likely to undergo liquifaction during a large earthquake. Across the Antioch Bridge, the rural character north of the river channel was replaced by shopping malls, housing tracts, and crowded highways in the communities of Antioch and Oakley.

On a dairy farm near Dutch Slough, east of Oakley, construction of 4,500 homes had been planned, but the farmer instead sold his 1,200 acres to the state. The land was six feet above sea level, which made it suitable for the largest freshwater tidal marsh restoration project in the Delta. We walked atop a levee, trying to visualize the dairy pasture as tule wetlands. "Come back in 10 or 20 years," John told us, "and you should see Valley Oaks and sycamores and native walnut trees grading out into grasslands and tidal marsh."

The restored wetlands, intermittently inundated by high tides, will provide important salmon habitat. Salmon spawn and are born much farther upstream in the mountains, but juvenile salmon, beginning their own treks from the Sierra Nevada to the sea, take advantage of tule marshes to pause, feed, and grow. As larger salmon, they will fare better when they finally enter the ocean.

Our final destination with John was a subdivision built below sea level in Oakley, one of the epicenters of the home foreclosure crisis in northern California. The levee that protects the houses meets 100-year FEMA standards, so homeowners are not required to buy flood insurance. With climate models forecasting a possible four-foot sea level rise by the end of the century, such faith in levees to protect houses seems a recipe for disaster. China's Yellow River provides lessons about the potential of disaster when living below water-level behind breakable levees.

John delivered us to our "mother ship," waiting near the confluence of the Sacramento and San Joaquin Rivers. The channel grew wider and wider as we sailed through Suisun Bay and, just before sunset, docked in Benicia, the major 19th century commercial port for produce shipped from Central Valley farms to Bay Area cities, until the railroads took over that job.

UPDATE 2020

A revised plan to move water beneath the Delta in a single tunnel was being explored in 2020. The Dutch Slough Tidal Restoration Project broke ground in 2018.

Strait to the Bay

The following morning, the 14th day of our cross-California trek, the very low tide left even the shallow-draft catamaran temporarily aground inside the Benicia marina. The Bay was nearly flat, the weather sunny and calm – perfect for paddling the 9.5 miles to Vallejo in kayaks. That morning will remain one of the highlight memories of our trek.

As we approached the Carquinez Strait, several massive ships came in beneath the bridges, generating impressive wakes that rocked our little boats even though we hugged the shore. Cars looked tiny on the freeway spans of the bridge high overhead. Paddling beneath them, we whooped out loud with exhilaration. We had arrived in San Francisco Bay, along with vital flowing water that had originated high in the Sierra Nevada.

On the far side of Carquinez Strait, we rounded a breakwater and entered Mare Strait. At that point, the rising tide began pushing us up the channel and we barely had to paddle to reach the Vallejo Municipal Marina, where we waited for the sailboats to catch up. There we took Marc Holmes on board and continued up the Napa River estuary.

Marc, tall and tanned with dark curly hair, is the Bay Restoration program director for the Bay Institute. He guided us a few miles up the Napa River to former Cargill salt ponds now being restored (38°09'N). Of the original 196,000 acres of tidal marshes from Suisun Marsh out through San Francisco Bay, eighty-six percent are gone. Opportunities for coastal wetlands restoration in the Napa-Sonoma marshes are enormous.

Approaching Carquinez Strait in kayaks, where water from the Sacramento-San Joaquin Delta enters San Francisco Bay

While we motored upriver, Marc pointed toward houses lining the northeast shoreline. "None of those residential areas could be permitted today, because they're built on the tidal margins of the Bay." The San Francisco Bay Conservation and Development Commission was established in 1965 to halt infill development, because, by then, one-third of San Francisco Bay had been filled or diked or drained. That followed a successful campaign by Save the Bay, an older organization than the Bay Institute, which functions much like the Chesapeake Bay Foundation to educate and advocate for this west coast mirror image; it is a striking coincidence that the nation has two gigantic coastal bays intersecting the 38th parallel.

Today, any flat land bordering the water, "used to be Bay," Marc said, "including San Francisco's financial district and Mission Bay." Communities along the edge had simply expanded by simply filling in shallow marshlands. "Plus, there were 48 garbage dumps on San Francisco Bay in marshes."

"Like 'Mt. Trashmore' in Alameda, where I grew up," Janet said. "I remember going with my dad over there. You just dumped whatever."

"Sure. Bring appliances down, refrigerators, back up your station wagon, and watch a bulldozer push the trash out into the marsh."

Everyone transferred to Marty's catamaran to motor into the salt pond complex. Tidal marsh had been converted there in the mid-1870s to grazing land and later to duck hunting clubs. The shallow wetlands were perfect for phenomenal numbers of migratory shorebirds and waterfowl. In 1956, Leslie Salt bought the properties and later sold them to Cargill. We were looking at salt evaporation ponds that each represented 1,000 acres of former wetlands. In 1994 the ponds were bought by the state and were now managed by the Coastal Conservancy.

"This complex worked by bringing Bay water in through a channel and circulating it through a series of 12 ponds," Marc said. "Eventually it was as concentrated as it would be. Decant off the brine solution and scoop up the salt crystals left sitting on these flats."

"Where did the levee go?" we wanted to know. Only a broken line of low vegetated clumps separated us from the nearest pond.

"Eighteen months ago, they made strategic breaches and razed them with bulldozers." They had used aerial photographs and old maps to select breach locations that matched old channels and excavated those inside the ponds before cutting the breaches.

"Look out there! I'm stunned to see all that cord grass." Marc was clearly excited at the pace of change. "This grass is brand new, in the last 18 months, without any planting; seeds traveled in on the tides. None had grown here in 100 years or more. This will start the natural cycle of a tidal marsh once again: detritus will fall, the elevation will gradually build up, and eventually we'll recover large pickleweed marshes."

"Take us through the rest of the developing food chain," we requested. "Now you've got the grass coming back, what else?

"First are invertebrates, they come back immediately. Then wading birds: Black- necked Stilts and American Avocets, and sometimes some shorter-legged waders as well, when the tide goes out. With inundation at high tide come Long-billed Curlews, some dabbling ducks, occasionally some diving ducks like Canvasbacks. In the winter time, these ponds are just lined shoulder to shoulder with migrating birds.

Thousands of ducks and other waterfowl benefit from salt pond restoration

"This is a very important fish nursery," Mark added, where many fish that spend their adult lives at sea come to reproduce. That is a key function of coastal estuaries everywhere, including the Yellow River delta and the *marismas* of the Guadalquivir River in Spain. "It is also the farthest west habitat for Delta Smelt, historically." The endangered Delta Smelt species is at the center of controversy over water diversions, because those powerful aqueduct pumps suck in and kill so many of the tiny fish.

"This is so great; you're seeing it right at the beginning," Marc said. "Come back here in five years and the cordgrass lines will extend across the pond and you'll have to get into kayaks to paddle up the channels.

217

Salt pond restoration was also underway in the south Bay, complicated by mercury in the bottom mud that washed down from mines before the ponds were diked and may be released into the Bay with restoration. Mercury becomes concentrated as bigger fish eat littler fish, leading to health warnings about humans eating too many fish caught in the Delta and Bay. This explained a startling sign (at first; the second and third lines just needed to be combined) that we had seen at Brannon Island, that read:

"Eating Fish Caught in the Delta
May Cause Pregnancy
or Slow Development Problems."

As at Chesapeake Bay, pollution has been a major problem for the west coast bay, culminating during the unrestrained runoff of sewage and industrial wastes directly into the water during and after World War II. The national Clean Water Act of 1977 provided the tool for controlling such "pipe" sources, and the odor and healthfulness of the Bay has improved tremendously. Today, as along the East Coast, the problem is how to control non-point pollution from millions of diffuse sources. Whenever it rains, storm drains bring a wave of petrochemicals and toxins washed off streets and yards; "Drains to the Bay" messages stenciled above sidewalk storm drains are as ubiquitous here as they were in Virginia.

Our goal, the next day, was to reach China Camp State Park and the neighboring national estuarine reserve. "At China Camp, you'll see the marsh as it existed 4,000 years ago when the Bay was formed," Marc told us. "Petaluma Marsh is the largest tidal marsh remnant. China Camp's is smaller, but one of the wonderful things there is a complete cross-section of the ecotone from open water bay to mudflat to tidal marsh to grassland to oak woodland. It's a precious little remnant."

A party atmosphere developed that afternoon – perfect weather, good company, beer, peanuts, and talking about good work being done restoring the Bay. After we dropped Marc off at the marina, the tide and weather was right for crossing, so we sailed over to the Petaluma marsh that Marc had described. It was nice to turn the motor noise off when Captain John put up the sails for a while.

Near the mouth of the Petaluma River, a few feet beyond a crumbling bank, the Sonoma Baylands marsh spread eastward, part of the San Pablo Bay National Wildlife Refuge. Just upriver was the start of the Petaluma marsh. After much anchoring, rope-tying, maneuvering, and discussions about currents, we settled in for the night at 38°07'N.

In the morning we sailed south across San Pablo Bay toward China Camp. From the middle of the Bay we could look back eastward toward the Carquinez Bridge and, looking southwest, see the distant skyscrapers of San Francisco. The catamaran had to be used for the shallow water landing, so we said goodbye to Captain John and David Martin. On the ride up onto the beach, we felt like Columbus coming ashore in the New World. China Camp beach is exactly on our latitude line (38°00'N). While we waited for our son, Ryan, to show up, we checked out the little museum and cafe.

China Camp's only resident, Frank Quan, is a spry, elderly man that runs a small cafe. In the late 19th century about 500 Chinese people, including his family, had lived at this shrimp camp, one of many around the Bay. Shrimp were sent all the way to China, as there was no local market in California. Two million pounds of shrimp were harvested each year until the Chinese bag nets were banned as a shrimping technique (a harvesting versus sustainability conflict that mirrors problems in the Mediterranean Sea). As with the historically over-harvested salmon in the Delta that persisted with smaller, yet still viable populations until recent years, Frank had still fished for shrimp in the Bay. Now the species is almost all gone, though, because they seek out the mixing zone to find less-salty water, and water diversions from the Delta mean less dilution down in the Bay. Historically, during the high flows of spring runoff, most of San Francisco Bay would be freshened. Frank no longer expected the shrimp population to last through his lifetime. The shrimp cocktail he served us in his restaurant came from Oregon.

On the bicycles again, thanks to Ryan, we continued by bicycle across the Marin peninsula through redwood groves, coastal scrub, and pastures where cattle grazed. At Olema, where we reached Highway 1, the ground was displaced as much as 21 feet on either side of the San Andreas fault by an earthquake on April 18, 1906, that is more famous for its

219

devastating effects on the City of San Francisco. Tomales Bay, a long arm of water a few miles north, is where the infamous San Andreas fault enters the ocean. By lunchtime we had left the North American tectonic plate for the Pacific plate and were at the Bear Valley headquarters of Point Reyes National Seashore (38°02'N).

Park superintendent, Don Neubacher, had attended the graduate School of Ecology at U.C. Davis with two of the founders of the Mono Lake Committee. He spread a map on a picnic table outside the park office and told us about the many projects they are working on to restore coastal marshes and reopen miles of streams to fish passage. Their biggest single effort was restoration of the 560-acre Giacomini marsh at the south end of Tomales Bay.

"It had been a dairy," Don said, "and they diked off the bay in the 1940s. Lagunitas Creek flows through there and is the one creek in this region that still has a pretty good run of Coho Salmon, with 15 to 20 percent of the remaining Coho in central California. You'll see 3-foot fish in 12 inches of water."

Don told us that the Giacomini project alone adds ten percent to central coast wetlands. "That gives you an idea how little is left," he said. "On October 25th, we'll open the last bit of levee and 50 to 60 percent of the land will flood at high tide. This will be transformed in a couple of years."

Camping that night in a volunteer campground not far from the headquarters, we heard the screechy calls of Spotted Owls from the nearby trees.

UPDATE 2020

Faced with sea level rise projections for the Pacific Coast and San Francisco Bay due to climate change, planners are designing sea walls for urban enclaves and hoping that tidal wetlands can help spread rising water, without being totally lost to inundation. The U.S. Geological Survey, National Academy of Sciences and other scientific bodies now estimate that the Pacific Ocean on the West coast will rise about two feet by 2050 and up to five feet by 2100. Recent accelerated melting of Greenland and Antarctic ice sheets may force those estimates to be revised upward. So far, little has been done to reduce humanity's greenhouse emissions.

Point Reyes and the Pacific Coast

The morning was overcast as we pedaled up the west shore of Tomales Bay, but we found the sunshine as we turned back south towards Drakes Beach. Our son, Ryan had driven there already and was photographing California Gulls on the beach when we arrived. We wondered if they were gulls born at Mono Lake, some of the 70 percent of that species in California that fly to the lake each spring to breed and raise chicks. The gulls spread along the entire state coastline during the winter, but these had perhaps chosen the most direct migratory route, following the 38th parallel as we had, to reach the ocean.

We inflated our two-person kayak and changed into wetsuits and life jackets. The bright yellow boat, our "rubber ducky," did not possess the grace of the wooden kayaks we relied on in the Delta, but only had to carry us two miles along the Bay.

We were novices at launching through waves. Three aborted tries sent us tumbling about on the beach, but Janet looked down the beach and said, "The waves look easier there." We carried the boat 100 yards, studied the wave pattern, then made a successful launch.

About halfway to the fish dock at Chimney Rock we heard the wonderfully mysterious calls of loons and then saw several of them dive and surface, calling repeatedly. Then from the beach came the guttural, vibrating "chonk" call of Elephant Seals. A half-dozen cigar-shaped seals were on the beach. Two reared up, displaying their protuberant noses to each other.

After kayaking across Drake's Bay, the authors walked across Point Reyes to the lighthouse where the 38th Parallel reaches the Pacific Coast

Ryan met us at the fish dock and we loaded gear into the car. Then we finished our trek across California by walking across Point Reyes.

The three miles still to go lay almost exactly along the 38th parallel. Cows studied us solemnly as we passed their pasture. A turnout provided a view straight over a cliff onto another gathering of Elephant Seals. We strolled up the road, feeling curiously reluctant about finishing the walk.

At the Point Reyes lighthouse parking lot, friends were waiting with food and champagne. Together, we walked to the lighthouse overlook (38°00'N; 123 01"W).

We stared at the ocean, trying to realize that there was no place left to walk. Off to the southwest, the profiles of the Farallones Islands were visible. Over the horizon, far away, the 38°N latitude was heading toward another landfall in Japan. Soon we would travel that direction.

Meanwhile, we took turns proposing toasts and speaking about the trek. A foghorn blew in the background, though it was another gloriously sunny, clear day, despite Point Reyes being the "windiest place on the Pacific Coast and the second foggiest place on the North American continent," according to the National Park Service. Before we left the overlook, an elderly couple that had been coming there for more than 30 years, told us that day had the best weather conditions they had ever seen at that spot. We were very lucky.

After finishing the cross-California trek, we soon returned for the first high-tide inundation of the Giacomini marsh at Point Reyes, joined migratory bird counters at the Golden Gate, and took a whale-watching cruise to the Farallones Islands – fitting points of closure for the California exploration.

About 500 people were at Point Reyes to help Don Neubacher and the National Seashore staff celebrate the re-watering of the Giacomini marsh. Led by a ranger, we walked across the valley to meet the tidewater, then steadily backed away as the leading edge kept pushing onshore. Humans were not the only spectators enjoying the event; rising water was evicting rodents from underground burrows and a host of Egrets and White-tailed Kites were patrolling the edge of the tide. Though just the beginning of a process that will restore a complex wetland ecosystem, it was exciting to see pastureland "going tidal" (as T-shirts on sale that day proclaimed) after so many years of planning, fund-raising, and preparatory work.

Hawks, Merlins, and Turkey Vultures soared above the Marin headlands north of the Golden Gate Bridge, as we learned about the Golden Gate Raptor Observatory from director, Allen Fish, and volunteer hawk counter Herb Brandt. Both had been with the program for over 25 years. Raptors migrating between North and South America seek out short water-crossings like the Golden Gate at the mouth of San Francisco Bay (37°49'N). Just as in Sicily, at the Strait of Messina, and at the mouth of Chesapeake Bay, birds are concentrated and most easily seen and counted where water and land funnel migratory birds.

The day before, 390 raptors had been tallied from Hawk Hill. Southward-traveling migrants arrive in late September, coming down the north coast and from inland California. In the spring, March through May, birds return from the south.

223

The Hawk Hill site had been part of the military bunker complex built to guard the Golden Gate. The land opened to the public in 1972 after the Golden Gate National Recreation Area was founded. The raptor count began after a "crazy lady," Allen said "began calling the California Academy every fall, saying she was seeing a Broad-winged Hawk every autumn by her house above San Francisco. No one had ever seen a Broad-winged Hawk here; it is an eastern woodland hawk that migrates down to the Amazon, so everybody passed the phone around and said, 'You talk to the crazy lady.'" As it turned out, there were many types of raptors, including over 100 Broad-wings every year. "It's the only spot in California to predictably see this species," Allen said. "Nobody had a clue that they were moving through here before that lady called."

A group of soaring Turkey Vultures caught his attention. Such interruptions are part of bird count conversations, when everyone stops and lifts binoculars to their eyes. "We love looking *down* on the backs of Turkey Vultures from this height," Allen said. "Oh, and here's an accipiter right in front of us, a Sharpie (Sharp-shinned Hawk). Look how quickly he's getting lift." The birds gained altitude on thermals rising over the south-facing hillside. "The 'money of migration' is lift," Allen said. Herb, a very experienced volunteer, explained that, "birds climb really high here, and then go out over the water and start dropping. They'll go out sometimes and come back, go higher, and finally take a line across."

The Raptor Observatory is funded by the Golden Gate National Parks Conservancy. Allen was proud of the Conservancy "for seeing there were ways of bringing people into a national park to do conservation work who didn't have to be PhDs, didn't need to be professional contractors; that they could create a community tradition." The hawk count program now has 300 volunteers.

"The 'ah ha!' moment occurred for me about five years into this," Allen said. "The truth is, this program offers therapy. There's a quality of daily calming and easing into urban life. People feel like they have a meaningful, useful, long term connection to the landscape. The joke is, 'Take two oak trees and call me in the morning.' What I feel like I'm 'selling' is more spiritual than scientific. The state of your inner nature is directly parallel to the outer nature where you

Near the Golden Gate bridge, Allen Fish told Janet about the annual counts of migrating hawks

spend time. That's why this is so much bigger than just monitoring birds of prey."

Our whale-watching boat chugged out of sunny San Francisco Bay through the Golden Gate and into a dense marine fog that persisted the entire 28 miles to the Farallones, the westernmost bits of California on our 38th parallel route (37°44'N; 123°02'W). A marine biology student, Jessie Beck, was with us to get her first close look at the islands before returning for 12 weeks of research there for the Point Reyes Bird Observatory.

After three hours over ocean swells, we finally approached the jagged "Devils Teeth" rocks, looming through the mist. A lighthouse stands on a small flat area of the 70-acre island, once staffed by light keepers and the Coast Guard who lived in two houses down below, until the light was automated. Researchers now occupy the houses, staying three to six-months to study seabirds, marine mammals, and sharks. The nesting bird colonies had been plundered for eggs during California's Gold Rush era and elephant and fur seals had been wiped out by the end of the 19th century, but now wildlife is protected within the Gulf of Farallones National Marine Sanctuary. Over 300,000 breeding seabirds use the islands; it is the largest colony in the lower 48 states. Only a few scientists are today allowed to set foot on the islands.

Jessie's interest in the Farallones began when she was young, on family trips to Point Reyes. "They seemed mystical and impossibly remote," she said. "My parents always asked us if we could see them in the distance and we would strain our eyes looking." Occasionally the fog would lift enough for glimpses of the island peaks, just at the horizon.

After 12 weeks studying the Farallones seabirds, Jessie told us that, "living on the islands was, first and foremost, long days of hard work, seven days a week. My life became consumed by the writhing mass of bird life. I woke up to the calls of Western Gulls, knew each nest along each pathway, spent the days monitoring those nests, ate dinner while watching baby Common Murres fledging, and fell asleep to Cassin's Auklets chirping. I began to recognize each bird's personality, their distinct calls and gestures to their partners. In the midst of a screamingly loud, guano-splattered chaos, I began to see the dense colonies as beautifully simpler microcosms of our own societies. It was the most rewarding work I'd ever done."

During our boat cruise, we saw sea lions and Humpbacked Whales cruising near the islands, but none of the Great White Sharks that are common there in other seasons. The boat trip itself was challenging enough to leave us in awe of researchers, like Jessie, who commit to harsh living conditions and isolation on the Farallones to better understand the needs of recovering species.

Our next landfall and final destination along the 38th parallel was Japan, another place of recovery and renewal.

UPDATE 2020
Salmon runs up Lagunitas Creek in 2018 brought the highest numbers in 12 years, but several additional habitat restoration projects in the area have been completed bringing hopes for future runs to improve even more.

Part V: Renewal

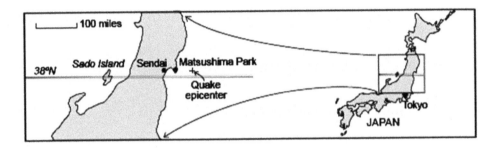

"The Toki [Japanese Crested Ibis] has the power to change people."
Ryugo Watanabe, General Manager of the Sado City Biodiversity Promotion Section (Fatakuchi, 2010)

Recovery on the Tsunami Coast

"Tsunami coming. Must go up!" Advice to an American tourist in Matsushima, Japan, from a local woman, on March 11, 2011, minutes after the Magnitude 9.0 Tohoku earthquake (Paul, 2011)

As we flew into Tokyo, the map displayed on the airplane-seat screen depicted not only landforms and our plane's location, but also physical features along the ocean bottom, including the Japan Trench, a dark blue gash extending north-south off the east coast of Honshu, Japan's largest island. That deep trench appears where the Pacific Plate bends downward and is overridden by continental plates. This collision zone on the Earth's crust explains why Japan is one of the world's most earthquake-prone nations.

On the morning of March 11, 2011, a massive earthquake occurred off the northeast coast of Honshu, with its epicenter at 38°19'N. The magnitude 9.0 quake, the fourth largest ever recorded, created a tsunami that reached the island's east coast in 30 minutes. A 30-foot wave struck Sendai, the largest city near the epicenter (38°15'N). Farther north, in Miyako, the wave funneled up a narrow coastal canyon that elevated its crest to an astounding 132 feet.

The tsunami caused much more damage than the quake itself or its many strong aftershocks. More than 21,000 people died and over half a million people were displaced. The quake also damaged nuclear reactors south of the line, most critically at Fukushima Daiichi, where several reactors lost cooling capability, experienced meltdowns and released radiation into the air and ocean. We arrived in Japan six months later, after

228

major recovery operations, food and water shortages, electricity outages, and the threat of radiation were addressed.

Japan is a land of contrasts: quietly speeding bullet trains, carefully tended gardens, clean well-kept cities, but also bright flashing lights, Kareoke bars and pinball machine gambling halls. It is a traveler's dream country of order, efficiency and politeness. The chaos brought by the quake and tsunami was incredibly upsetting for such an orderly people.

Our flight arrived in Tokyo and, before leaving there to head north to the 38th parallel, we visited the Ministry of Environment national headquarters to learn about a plan for a new national park on the coastline impacted by the tsunami. Several existing national, prefectural, and local parks will be incorporated into the Sanriku Fukko ("Recovery") National Park. "Sanriku" is the name given to the northeast coast of the Tohoku region (a designation something like "New England"), which includes four prefectures (equivalent to our states).

The humidity and heat of central Tokyo reminded us of similar weather we had experienced in Korea. At the Nature Conservation Bureau offices, we located Shinjiro Sasaki, an assistant director of National Parks. The office space in a large room was a warren of side-by-side desks, dimly lit by the windows and lights of computer screens, because overhead lights were turned off. The air conditioning was also off and everyone wore short-sleeved shirts. That summer, the national government asked the nation to take such energy-saving measures to reduce the likelihood of power black-outs, because all but 15 of the nation's 54 nuclear reactors were shut down after the earthquake. Mr. Shinjiro gathered several other administrators and planners into a grand meeting room with large windows overlooking downtown Tokyo (we were on the 26th floor). The Director of National Parks, Tetsuro Uesugi, and the Director General of the Nature Conservation Bureau, Tsuneo Watanabe joined the meeting.

Over tea, they told us how plans were progressing for the new national park. A vision planning meeting had occurred just four days earlier, but much work remained to begin the project, including forging contacts with the communities struck by the tsunami.

"What prompted this idea for a new park?" Janet asked.

"At first, we were overwhelmed by the earthquake disaster," our translator summarized, after Mr. Watanabe and Mr. Sasaki answered. "We were here in Tokyo, but wanted to help those people because the National Parks there are a key to their local economy and the most significant features in much of that area. We have staff in several parks there now, but hope to expand with this new concept for a national park." As we would see in action, later that week, Japan's national park rangers implement national conservation policy, not only on park lands, but in communities and regions where environmental projects are underway. This project will increase tourism, provide employment to locals as park guides, and honor the victims of the tsunami. A new trail will be built along 200 miles of the coast, to not only connect the parks, but also provide escape routes to higher ground during future tsunamis. They will also aid the local fishing industry by promoting eco-tourism on fishing boats.

Funding was an unresolved issue – they all chuckled and rolled their eyes when asked about the project budget – but Director-General Watanabe said that special reconstruction funds established after the quake by the legislature might be available, as well as private funding. Though still only at the vision stage, he emphasized that this Sanriku Fukko National Park was now his agency's most important effort. Beach cleanup of tsunami debris had begun that summer directed by the Ministry of the Environment.

We would visit Matsushima Bay, in the southern portion of the proposed new park, where the nation's first *natural* park was created in 1931, renowned as one of Japan's three most beautiful sites. "Natural parks" are operated by local prefectures rather than the national parks division; one challenge for the Tokyo office will be coordinating with other levels of government affected by the new park project.

The Tohoku bullet train from Tokyo to Sendai was so smooth and silent, that it was hard to believe we were going 200 miles per hour. A scheduled stop was Fukushima City, 39 miles inland from the Daiichi nuclear power plants on the coast. Though the city is well outside the 12-mile mandatory evacuation zone and separated from the coast by a low mountain range, some radiation had drifted that far. Efforts continued there to scrub roofs and remove radioactive

Sendai tsunami debris

materials from private houses, parklands and meeting venues, with a particular emphasis devoted to households with children, all aimed at reassuring local citizens about conditions in that city.

The tsunami wave that struck the Daiichi nuclear facility on March 11 was 46 feet tall when it swept over a 20-foot seawall meant to protect the power plants. The meltdown of three reactors that resulted was eventually declared a "Level 7 event," the highest on the International Nuclear Event Scale (equivalent to the Chernobyl meltdown in 1986). The mandatory 12-mile evacuation zone imposed by the Japanese Government affected over 7,000 businesses and over 100,000 residents. Some of those unfortunate people were still living in shelters six months later, while the situation at the nuclear reactors themselves remained extremely serious. There was plutonium pollution at the power plants and radioactive water still leaked into the Pacific Ocean.

We were glad that stops are brief on the Shinkansen bullet trains, and soon sped away from Fukushima City towards Miyagi Prefecture's capital city of Sendai, sixty miles farther north. Within a half-hour of the March 11 earthquake, a 30-

foot high tsunami wave had swept across Sendai's airport, carrying cars, trucks and tons of debris inland. Videos of that dramatic scene, with people gathered on the airport terminal roof as the wave swept inexorably onshore, were televised again and again during news coverage of the earthquake. In Natori, the airport city, the tsunami killed 1,019 people and destroyed or damaged more than 8,000 homes.

The tsunami did not touch Central Sendai, which is on higher ground and had only minor quake damage, but for weeks afterward, more than 1 million people had no electricity or gas and were isolated by damage to roads, train tracks, and the airport. Air service resumed a month later and trains began running on April 29. After six months, hotels for tourists and business travelers were open and life was more normal; tourists were trickling back.

Sendai is west of Matsushima Bay at the south end of the Tohoku region. A ten-mile train-ride from downtown brought us to the local port for fishing vessels and tour boats. Twenty-two people had died near the port, a relatively smaller number because the force and height of the tsunami was dampened by offshore islands. There are 260 small islands in Matsushima Bay, most of them crowned by pine trees ("*matsu*" means "pine" and "*shima*"is "island"). The first tsunami damage we saw was a damaged seawall and sidewalk along the waterfront. The woman who sold us our tour boat tickets came outside to show us where the wave had crested four feet up the wall of the terminal building as she and others escaped to the roof.

Before boarding the boat, we bought one of the souvenir T-shirts for sale everywhere in the region that proclaim: "*Ganbarou Tohoku!*" (Let's Do Our Best, Tohoku) or "*Ganbatte Nippon!*" (You Can Do It, Japan), expressions of determination and encouragement.

We had the morning tour to ourselves for the 50-minute trip that wound among the scenic islands and finished at the town of Matsushima. The famous bay remains a magical place, worthy of its reputation for beauty. The larger islands are inhabited by fishermen and seaweed farmers. A small one is named "Horse Set Free Island," because samurai used to turn old horses loose there to live out their days. Most of the islands are even smaller rock stacks, intricately sculpted by the water, topped by elegant-looking Japanese Red Pines

(*Pinus densiflora*), a symbol of longevity in Japan. Here and there, trees had been broken by the force of the tsunami or killed by inundation with salt water. Several wrecked boats still lay on their sides on beaches, six months after the earthquake.

At the town of Matsushima, population 16,000, we located the headquarters of Matsushima Bay Natural Park. The office building is only a few feet from the bayshore and had been badly damaged by the tsunami, but since rebuilt. No one spoke English in the small office, but one of the park administrators, Hiroya Miura, kindly led us on foot for several blocks to the Tourist Information kiosk at the train station, where a woman could translate. Mr. Miura was aware of the new Recovery Park being proposed in Tokyo.

When asked his opinion of that effort, after a long pause he said, "It will be difficult to coordinate so many different levels of national and local park operations," then added, "but, yes, it could be a way to increase tourism and provide jobs." The national park planners in Tokyo have much work to do.

A bridge to one of the closest islands offshore, just behind park headquarters, had been completely destroyed by the tsunami and not yet rebuilt. On the September day we visited, waterfront cafes bustled with Japanese tourists having lunch. The cafes were part of a line of businesses that is separated from the bay by only a 2-lane road and a low seawall. The ground floors of all those buildings were flooded when the relatively small tsunami wave arrived (only about 5 feet high). That left a mess to cleanup, but did relatively little structural damage. The shaking itself caused more damage in this village, destroying 200 homes, including our English translator's (it had already been rebuilt). Snow had been falling when the quake hit, and power was knocked out, so keeping warm at night was added to the many challenges residents and travelers faced right after the quake.

A long graceful bridge over to Fukuurajima Island survived, though damaged tiles on the approach walkway were being repaired. A chorus of strange frog calls provided the background serenade to our hour-long hike around the lush island. Labels identified many of the 300 species of trees and grasses found there. Coves and beaches reached along the main trail appeared pristine, but a little-used side path

233

Objects, including shoes, were left by the March, 2011, tsunami in a cove in Matsushima Bay

dropped down to a remote cove that was still covered by tsunami debris. Among the pieces of lumber and assorted flotsam, were a thermos, light bulbs, a wheel from a child's scooter, and two shoes of different colors and sizes, leaving us with unanswerable questions and a sudden sense of losses that occurred.

This region has a long history of both earthquakes and tsunamis. Sendai was swept by a massive tsunami 1,142 years before, following an earthquake of unknown size, according to written records and layers of sand deposited in inland soils. In 1896, the Meiji-Sanriku earthquake, magnitude 7.2, generated a tsunami that destroyed about 9,000 homes and killed at least 22,000 people farther north along the Sanriku coast. During the 20th century, no earthquakes there reached magnitude 8, until March 11, 2011.

Two months later, the U.S. Geological Survey had counted 525 aftershocks exceeding magnitude 5.0, including five above magnitude 7. On July 31, 2011, the region felt a 6.4 magnitude aftershock, at a time when the disaster-ravaged population had just endured a torrential summer rainstorm that dumped over 26 inches in three days and led to flooding evacuations in the Fukushima and Niigata prefectures.

During the night after we visited Matsushima, an earthquake shook our beds in Sendai, waking us up near midnight. It was only magnitude 4.7, a minor temblor, hardly noticeable in that part of the world.

We departed that morning on another train, heading west again along the 38th parallel through the island's interior. Though the local park official foresaw many obstacles to the plan being considered at national park headquarters, in the next few days we would see an impressive collaboration of farmers, rangers, scientists, and citizens to adjust a way of life, all to rescue the endangered Japanese Crested Ibis. If such determination and energy represent the national norm, then recovery and renewal along the Sanriku coast will almost certainly also be achieved.

UPDATE 2020

Inauguration ceremonies for the Sanriku Restoration National Park were held on May 25, 2013. Two years later, the Ministry of the Environment added the former Minami Sanriku Kinkasan Quasi-National Park. The park will eventually include Kesennuma Prefectural Natural Park, Kenjōsan Mangokuura Prefectural Natural Park, and Matsushima Prefectural Natural Park.

Nobiru Station was among a handful of structures that remained standing after the 2011 tsunami swept through. Part of the former train station is now a museum and community space memorializing the tsunami.

Japanese Crested Ibis: It Takes a Village

Halfway across the volcanic interior of Honshu Island, we switched from a train to a bus because heavy rains that summer had damaged tracks along our route to the west coast city of Niigata (37°55'N). With Japan's dense population and limited space, the forested mountains of the interior contrasted with land otherwise intensely cultivated or inhabited. Ubiquitous rice fields, turning golden-yellow just before the harvest, were strikingly beautiful where they met the edges of dark-green forests. The border zone between mountain foothills and arable land is called *satoyama,* and is the critically important habitat type for the endangered Japanese Crested Ibis. Called "Toki" by the Japanese, the large pale-pink bird is the avian symbol of the nation; its scientific genus and species names reflect that special significance: *Nipponia nippon ("Nippon"* means "Japan").

Niigata is a large port city serving the west coast of Honshu Island, where we would catch a ferry at dawn to Sado Island. Professor Tsuneo Sekijima of Niigata University met us the night we arrived and, before hosting a traditional Japanese dinner, introduced us to the Toki's complex and compelling story.

Once common across Japan, the large birds thrived for thousands of years in the bio-cultural landscape humans created on the islands, where rice paddies came to dominate much of the landscape. With wingspans up to 4.5 feet and a very long down-curved bill, Toki are dramatic looking birds. In the past, they roosted in trees close to the paddies where they fed, the same locations where people lived and worked,

Japanese Crested Ibis, called "toki," fly again on Sado Island, due to a conservation program that brought them back from the brink of extinction

and the birds became a familiar, precious part of national life.

Though their overall size is similar to a Great Egret, the Crested Ibis have shorter legs and need to forage in water less than four inches deep. In the mud and shallow water of rice paddies, irrigation canals, and natural wetlands, the sensitive tips of their long beaks probe for the touch of insects, worms, and fish.

There had always been hunting pressure on the birds, to eat and to collect their beautiful pink feathers, but the decline in the later decades of the 20th century coincided with changes in rice culture. Herbicides and pesticides destroyed the food chain that fed the Toki. Irrigation ditches moving river water to paddy fields were turned into concrete-lined channels with gates to control floods and allow farmers to drain their fields. Fish, especially a slim loach that is the key food for the birds, could no longer swim up from flowing channels into the paddies to spawn. To facilitate the use of heavy tractors (in place of livestock), rice fields were dried up before the harvest, which shut down biological productivity each year. Trees the Tokis need for roosting and nesting were cut for fuel, especially during the two world wars. By 1971, Sado Island, bisected by the 38th parallel, became the only remaining place where Toki flew in the wild in Japan. They had also been disappearing across their extended range outside Japan, including eastern Russia, eastern China, and the Korean Peninsula.

In 1981, in a desperate effort to save the species, Japan's last five wild Toki were caught and captive-breeding efforts began. That same year, a remnant population (only seven birds) was rediscovered in China, where the species had also been considered extinct. The Chinese began a successful captive-breeding program and, in 1999, gave Japan two Toki, then sent one more in 2000, and two in 2007. The last of Japan's once wild birds died of old age in 2003.

There is, clearly, no point in breeding and releasing birds back into the same environment that caused their extinction in the first place. The story of the Japanese Crested Ibis is unique – and was an inspiring way to end our around-the-world travels – because the urban and rural residents of Sado embraced key changes to their island lives and livelihoods. Dr. Sekijima's research to assess the habitat needs of Toki are just one part of an unusual cooperation between university, government, environmental organization, and local people, including children and, critically important, the certified eco-farmers of Sado who now produce "Toki Brand" rice.

We were on the first ferry from Niigata, leaving at dawn for the two-hour trip to Ryotsu, the port city of Sado Island (38 05'N). A ranger wearing the yellow uniform shirt of the Japanese National Park system waited there to meet us. The large island (over 200,000 acres) is mountainous, with its tallest peak at 3,845 feet, and home to 62,000 people, with about 25 percent of the population involved in rice farming.

We spent the rest of that first day with National Park Ranger Kei Osada. Gray-streaked hair belies his young appearance and energy. In a 15-year career as a ranger, he has worked elsewhere, including the Sanriku coast and at the Tokyo headquarters, but came to Sado Island just 14 months earlier to work at the Crested Ibis Conservation Center. "The role of a ranger in Japan is changing," he told us. "Now we are encouraged to be involved in community conservation efforts as well as taking care of parkland." The day he had mapped out for us was the perfect illustration of that new role.

Our first stop was at a community work-project where local farmers, members of the Biodiversity Conservation Network of Niigata, Sado City officials, and park staff had gathered to remove large-mouth bass and blue gill perch from a farmer's pond. The non-native species were introduced 30 years ago from the United States for sports fishing, but they devour loach, the small native fish, a critical part of the Toki diet. The farmer, Mr. Zutsumi, who owned the pond was there, and we met a Sado City resident who cooks at a noodle restaurant, one of a dozen men and women volunteers. The pond had been drained, so that bass and perch were visible, flopping around in shallow pools. Under the eyes of local TV cameras and reporters, the fish were snagged with nets on long poles, exciting a flurry of thrashing fish and flying mud as men in waders netted and heaped fish into boats and rafts. Thousands of fish were caught that day; they would be distributed to locals to eat. Ranger Osada was in the middle of the action, scooping up fish, then being interviewed for local television.

Signs along the edge of the surrounding rice paddies showed that they met requirements for Toki Brand rice. To receive Sado City's certification as an eco-farm, methods must be adopted that nurture living things, such as dirt-lined ditches (rather than concrete) bringing irrigation water to the

239

Rice farmers, environmentalists, Sado City officials, and national park rangers worked together to remove non-native fish to help the native food supply for Japanese Crested Ibis on Sado Island

paddies, where loach and insects can thrive; establishing small wetland habitats close by, called "biotopes," that are rich in food for the endangered birds; building fish ladders or ramps to connect ditches with paddies; and flooding fields in the winter, after the rice is harvested. Eco-farmers must also use levels of pesticides and chemical fertilizers at least 50 percent below levels typically used in that region, and report the results of annual biological surveys to the city.

Certified farmers then can market "Toki Brand" rice in bags that include a prominent picture of the large pinkish bird. Their rice is priced higher than that grown elsewhere, but the whole nation is aware of these efforts to save the species and the premium rice sells successfully in major Japanese supermarket chains. Until the program was established and Toki Brand Rice proved itself marketable, "many farmers were predominantly negative about the Crested Ibis project, saying "the release of Crested Ibises will fail" or it was "an effort irrelevant to agriculture." (Sekijima, et. al, 2010, 140). But after the first release of birds, in 2008, the number of participating eco-farms tripled, to 695 by 2010.

The city's eco-certification guidelines rely on research findings by Dr. Sekijima and his cadre of graduate students, and also the work of Dr. Yoishi Kawaguchi, a river restoration specialist from Tokushima University. He had joined us for dinner in Niigata and, on our second day on Sado, showed us field examples of his research. His studies clarified the irrigation system effects on Toki and their prey and the benefits of biotopes, and he designed the fish ladders being used.

Cleaning out every single fish in the farm pond was going to take all day. Meanwhile, Ranger Osada drove us to join a group of 20 elementary school children who had enrolled with Sado City's environmental education program on a wildlife survey, catching creatures in the Ishidu River. Though the channel was lined with concrete, the river was still rich with small fish, insects and even fuzzy-clawed crabs, which the kids captured in clear plastic containers to be identified, drawn on art paper, and finally returned to the river. This was part of the local effort to improve conditions for Ibis by focusing on improvements for "the small things," and to involve *all* members of the community in that effort. Sado City also has a Kids' Paddy Project, where farm kids and children from urban areas grow rice in pesticide-free rice fields, cultivate biotope habitats, and study the response of living things.

Kei then drove us up a winding mountain road as the sky grew darker and a few raindrops began to fall. On Takano Takeshi's farm, built on terraces in a canyon skirted by oak, beech, and pine trees, we joined twelve students from Wasada University who were energetically stirring mud in a rice paddy to enhance oxygen flow in that biotope. They had volunteered to come from their homes near Tokyo for four days; the concern for Toki is national and not only felt on Sado Island. Mr. Takeshi is the son of a local farmer who became famous for hand-raising the offspring of Japan's last wild Toki, after it was born in captivity. One of the earliest memories he can

241

recall is of Tokis perched in the trees outside his home. Ever since, he told us, they have remained special to him.

We had yet to see an actual Toki, though all of these projects ultimately intended to benefit the birds. After a snack with the college students that included "Toki milk" in a carton shaped like a fat bird (even Sado Island's dairy farmers are involved), Ranger Osada took us to the Conservation Center, where he spends most of his days, and we finally got a close look at Japanese Crested Ibis. As we came through a gate in a high fence, Kei raised his voice and spoke to the birds so they would not be startled. The cage held 18 birds who were moved to this training cage from eight other rearing and breeding cages. In the big training cage they can fly, socialize, and get ready for their transition to the wild. Each bird eats 40 loaches a day (raised at a fish farm) plus special meal pellet supplements. Food slides down to feeding ponds inside the cage from a nearby building, so that human contact is minimized. All the cages are monitored with video cameras serving a bank of video screens in Kei's outer office (similar to the center for Iberian Lynx we had visited in Spain). Despite the security systems, in March, 2010, a marten (a member of the weasel family) broke into a cage and killed eight birds. After that setback, Ranger Osada transferred from Tokyo to Sado Island to increase the Conservation Center's staff levels.

About 200,000 tourists come each year to the Toki Forest Park visitor center, outside the rearing area, to learn about and see the birds in cages. That operation is partially supported by GIAHS, the "Globally Important Agricultural Heritage System" of the United Nations.

The Conservation Center's first release of birds came in 2008. For the first time in 27 years, ten Toki flew wild over Japan. That day, fifth and sixth graders from Gouya Elementary School in Sado City sang the school's song for the Crested Ibis, to "encourage the birds to fly high in the sky." Marking the significance of that moment to the nation, the Japanese Imperial family was represented by Prince and Princess Akishinomiya. Though the first releases were from individual boxes, in each year since, a less disturbing "soft release" technique has been used: simply opening the doors of the training cage and letting the birds decide when they are ready to leave. In 2010, some birds took five days to depart.

Another 19 Toki were released in 2009, 13 in 2010 (the year of the marten attack), and Kei would release 18, two weeks after our 2011 visit, including eleven males, aged one to five and seven females aged one to three. The timing, late in September, coincides with the end of rice harvest and is when the wild birds naturally gather into bigger flocks for the winter. The goal is for 60 Toki to be flying wild by 2015.

The relatively small wild population is backed up by 132 Toki that now live in the Conservation Center (plus 41 more in three zoos in Japan). Though there have been three earlier releases, and some of those free birds have paired up, built nests, and laid eggs, none of their eggs have yet been successfully hatched.

China now has over 1,800 *Nipponia nippon*, with about 200 in the wild. Reproduction there has been more successful, perhaps because they began their breeding program with wild birds, instead of birds raised in captivity. Japan hopes someday to add more birds from China, in part to widen the genetic "bottleneck" that occurs when a whole population descends from just a few individuals.

Our busy day continued into the evening, as Ranger Osada escorted us to a dinner with the Wasada University students and local residents. The students enthusiastically practiced their English, telling us about their four-day adventure on Sado. They were curious about why we were there, asked what we had seen elsewhere on the 38th parallel, and wondered if they could read our book in Japan once it was published. Some of them, not surprisingly, were majoring in biology or environmental studies, but several were economics, sociology, or psychology majors. Perhaps, after that volunteer experience, they will find ways to build an environmental focus into those university majors. The
evening culminated with a performance of traditional Sado Island dancers and drummers.

Seeing such a rare bird up close, in cages, had been a pleasure, but early next morning, we saw wild Toki flying free. That was an exceptional thrill.

Our escort was Dr. Chihiro Endo, a soft-spoken post-doctorate student who looked too young to already have her

Children from Sado Island conduct wildlife surveys to learn about the river ecosystems and the food supply for Japanese Crested Ibis

PhD. The Toki's morning flight from groves of trees where they roost overnight is fairly predictable. Right on schedule, as the sun rose above the trees, a flock of six birds suddenly appeared. We saw several other flocks at a distance, heading for feeding grounds, and then three birds flew low over our van, before settling in trees, which gave us a great look through Dr. Endo's spotting scope.

"Toki need a certain landscape of rich rice fields with trees alongside to rest and preen in," Dr. Endo explained, reinforcing the importance of *satoyama*. "Just because there is a lot of food available in a place does not mean they will come." In fact, when rice nears maturity, as it was during our visit, the birds cannot forage among the tall stalks.

In the heat of mid-day, the birds perch high in trees to rest and preen. With no flight activity to watch, Dr. Endo switched to surveys of their prey – insects, frogs and loach -- in a fallow rice field. Unexpectedly, a single Toki flew over, delighting us all.

One of the volunteer bird surveyors that morning was a local carpenter who found a couple of Toki feathers in the mud (whitish, at that time of year, but with a pink shaft). He also showed us footprints made by the Toki in the mud and another set made by a Japanese raccoon. That gifted amateur naturalist gave us a set of excellent Toki photographs he had taken. We asked if we might use one in our book, giving him credit; he gave us permission, but preferred anonymity.

Sado Islanders have committed to making the changes necessary to their human system to sustain the special birds that inhabit their island. The mayor of Sado City, Koichiro Takano, is an enthusiastic promoter of this effort (Kei said that people consider their mayor himself close to being a Toki, because he is so involved). Mayor Takono describes the effort as, "a direct response to the biodiversity crisis that confronts humanity. Our efforts to bring back the Toki through nurturing the small creatures on which it thrives is an attempt to restore the environment that is worthy of worldwide attention." (Takano, 2010, 7).

On September 27, 2011, two weeks after our visit, the door of the Conservation Center's training cage was opened at dawn.

Within three hours, four male and three female ibises had flown away. Once the more cautious birds joined the others, 51 Japanese Crested Ibis flew, once again, over the island where the species had once gone extinct.

The recovering population of Japanese Crested Ibis is a symbol of hope and renewal and a model of sustainability for all of Japan and the world. The people of that small island, supported by the rest of the nation, are showing that environmental renewal can revitalize an economy, while that improved economy, in turn, helps fund restoration efforts. The endangered birds require a healthy water environment, as do the people of Sado Island...

...as do the people of the world.

UPDATE 2020

It has been 12 years since the first release of ten Japanese Crested Ibis. Since 2008, each year at the spring and autumn bird releases, naturally bred birds have been included, bringing current Ibis figures in the wild to over 290.

The Japanese Ministry of the Environment had aimed to establish 220 birds in the entire Sado Island region by the year 2020, but achieved that goal in 2018.

The Passion of Place

"What is a line but a dream and the will to follow it? What is an adventure but a challenge to oneself to go a little further off the beaten path?"   From Janet's journal

Circling the globe along the 38th parallel is a 19,600-mile journey. On the ground in Europe, Asia, and North America, the latitude line traverses about 10,800 miles; the Pacific Ocean is 5,300 miles across and the Atlantic about 3,500 miles.

China and the United States both span similar distances along the 38° line. The longitude is 119°E0020at the Yellow River delta on China's east coast; at the western edge of that country we were at 75°E, having crossed 44 degrees of latitude. It is striking that the longitude numbers for the United States are nearly identical, though measured in the opposite direction from the zero point of the Greenwich Prime Meridian: Mono Lake is at 119°W (the Pacific coast at 123°W), and the Atlantic coast at Assateague Island is 75°W. Move away from the latitude line to the north or south, in either nation, and national boundaries no longer coincide so closely; this is an intriguing 38th parallel phenomenon. Both nations have moist eastern zones and arid western deserts, with differences in the climate and landscape explained by America's western border being an ocean, while China borders the mountainous continental interior of Central Asia.

These facts answer a question many people asked us: was the weather the same everywhere along the latitude line? It was not, of course. Nevada's desert is very different from the prairie in Kansas, which is unlike the Missouri woodlands or the green pastures of Kentucky. It was helpful to personally experience the transition from humid east to dry west in the United States, and actually feel the abrupt change near the 100th Meridian, where hair stopped curling due to Eastern humidity and Janet suddenly needed lip balm again.

When we reached the Berry's farm in Kentucky, Tanya, Wendell's wife, speculated that orientation to the sky and the angle of light might make people most comfortable elsewhere along their "home latitude." That particular relationship to the sky, which changes when traveling north or south, can become very noticeable if made in a fast jump by air. Holding to our 38th parallel course, different altitudes and distance from the ocean instead produced the most noticeable landscape and weather changes.

Summer rains that arrive with the Asian monsoon, like those we experienced in Korea and saw evidence of in the landscapes across China and Japan, were the most striking weather contrast with California's climate, where the Mediterranean pattern confines most precipitation to the winter. Korea and Japan's summer-time humidity can be overpowering to travelers from the western United States. The climate became much more familiar in western Turkey, Greece, Italy, Spain and Portugal.

We approached the California segment differently than elsewhere in the world. The 220- mile span across our home state seemed vast when walked at two miles per hour, sailed at four miles per hour, and bicycled, at times, up steep hills at a "speed" easily exceeded by a cruising butterfly. During those seventeen days we actually meandered 350 miles, 75 of them on foot, 168 on bicycles, and 107 miles on boats. Crossing slowly, with time to look at things closely and experience them directly, provided new lessons for the two of us, who were born in the state, and worked in many of its regions. Following part of the vast watershed that sends Sierra Nevada snow-melt out to the ocean through San Francisco Bay, it became clear that mountain winters we endure provide vital water to everything that lives downhill and that most of the water problems facing California have parallels around the world along the 38th parallel.

In her journal, Janet wrote: "I learned more about friends who shared our dream and encouraged us and helped us to make it happen. Our sons were there when we needed them with cold drinks and smiles and care for our empty house. I learned about my husband, his passion and determination, his doggedness and willpower. And I learned about myself. When I got tossed around in the ocean three times trying to get onto the kayak, I knew Dave yearned to paddle across to the Point Reyes lighthouse, to complete the California trek in a grand way. I am proud that I wiped salt water out of my eyes, smiled at him and went for it that fourth time."

The Asian countries of Japan, Korea, China, Turkmenistan, and Turkey took us farthest from familiar cultures and languages. Reaching the Korean DMZ introduced us to high school students that aim to shape the future of that environment, which is ironically recovering behind barbed wire barriers of warring states. Yet, it was distressing to watch South Korea's national river "restoration" project destroy riparian wetlands.

Cranes fly where China's Yellow River empties into the sea on the 38th parallel, but as we crossed that nation, burgeoning with economic growth, the new "holy crane" of China appeared to be construction cranes, whose gangly towers and "beaks" were at work almost everywhere. New coal plants are being rapidly built in China, yet the nation also shows laudable concern for the climate change problem they aggravate, as it aggressively pursues alternative energies and water-saving procedures for processing coal.

China's decision to build dams on its southern rivers and move water to the north to serve growing thirsts, made us ask the same questions we have asked about Los Angeles and that Glennon Zelch asked about Las Vegas: why not decide the ultimate size of a city according to its available, regional water resources? When will growth stabilize, as it inevitably must? And what quality of life and environment will humans choose, if they give considered thought to the long-term?

Many of the water problems we encountered remain unresolved. No one seems to be effectively devoted to reducing river diversions within Turkmenistan to help the Aral Sea. Though international voices speak out against Turkey's plan to flood ancient Hasankeyf, that government still continues to build Ilisu dam on the Tigris River. Tensions between Turkey and nations that share the Tigris and Euphrates, downstream, heighten concerns about peace in the Middle East. The environment of the Mediterranean Sea seemed "tired" to us, with too few birds and sea mammals, as if worn out by centuries of human exploitation. Intensive olive culture in Spain and Portugal demands more irrigation water in nations already pushed to the limits for water. Despite decades of efforts to clean up Chesapeake Bay and restore oyster and crab populations, growing human numbers outpace those efforts. Mountaintop Removal coal mines still push mining "wastes" into the creeks of West Virginia and Kentucky. The Ogallala aquifer that supports so much grain and meat production in the center of the United States is over-exploited. Mid-west farmers keep pumping water out of the ground faster than it can be replenished and sending it around the world as corn syrup products, but their life-style and economy must inevitably change as pumps go dry. Las Vegas maneuvers to acquire the groundwater of Northern Nevada to serve a never-ending thirst for growth.

And yet...

We found inspiration in the work of Chooney Kim and her co-workers with KFEM in South Korea. We were amazed by the dedication of Ran Liping and Green Camel Bell in western China, and stand in awe of Anna Giordano's brave, successful efforts to change a Sicilian culture that used to slaughter migrating hawks as they congregated at the narrow water crossing to mainland Italy from Messina. Spain's Iberian Lynx breeding and recovery program in the Guadalquivir River watershed is impressive, as is the uphill battle for the environment by the Quercus organization in Portugal. Environmental educators of the Chesapeake Bay Foundation must feel satisfying hope at the end of each day giving school children from the upper watershed their first experiences with crabs and oysters.

Local and regional population growth exacerbates every issue we came across, even in nations where national population levels have stabilized. The 38th parallel nations are unusual, because birth rates in almost every nation we visited have dropped towards, or below, replacement levels. Afghanistan and Tajikistan, two countries we avoided because of safety concerns, have the highest population growth rates among the 38th parallel nations. The other nations have benefited from what economists call "the demographic bonus." When countries move quickly to smaller families, growth in the number of young dependents – those who need nurturing and educating – declines relative to the number of working adults. As Lester Brown added in his book aimed at saving civilization, *Plan B 4.0*, the bonus allows productivity, savings and investment, and economic growth to increase. "Japan, which cut its population growth in half between 1951 and 1958, was one of the first countries to benefit from the demographic bonus. South Korea... followed, and more recently China [has] benefited from earlier sharp reductions in birth rates. This effect lasts for only a few decades, but it is usually enough to launch a country into the modern era. Indeed, except for a few oil-rich countries, no developing country has successfully modernized without slowing population growth" (Brown, 2009, 185).

Despite considerable efforts to plan our days, along the way we had unexpected encounters with people eager to share their lives and stories. Conversations often began: "Where are you from?" After we explained how and why we came to that place, stories would emerge. If not personal information, often we heard about someone else we should talk to.

In that serendipitous way, we met the county clerk on the west bank of the Mississippi River who kept a little museum open and eagerly talked about his former life as a riverboat pilot. There was a man outside Louisville, Kentucky, with a passion for local history and an amazing collection of old hand tools and other artifacts at a local history museum that we only found because a heavy rainstorm forced us off the main highway in search of shelter.

Passionate people were the rays of hope shining around the world.

We learned about Leo Dry, who established the Pioneer Forest in Missouri. He declared there was a better way to grow hardwood forests and set out to demonstrate his theories on his own land, changing the accepted paradigm for forestry in that region.

In the San Luis Valley of Colorado, Paul Robertson, with the Nature Conservancy, sat us down in a cafe and said, "This is how the story began..." and ran through the entire groundwater and land protection efforts of that valley.

Mark Rude, with the Kansas groundwater district, works to protect the Ogallala aquifer near Garden City, Kansas. He was a fantastic source of information; a thoughtful man who could see and explain the big picture.

So many good people were doing good work along the 38th parallel. Those people were unanticipated discoveries during travels that, when we first conceived this project, focused more were exploring the landscape along our route. Many of the people, like the staff of the Pioneer Forest or the local activist and hydrologists working to clean up uranium tailings in Moab, Utah, had been at their locations for many years and had no plans to move away in search of "greener pastures." They recognized the value of settling in, building their knowledge of the land, and caring for the local environment and their local society. Within each of these people there was a passion about place. We could see it in their eyes; we will always remember those young foresters in the Pioneer Forest and how proud they were.

Americans often relocate too often to build such strong feelings of responsibility for their homes (despite the life-work of poet and writer Wendell Barry which eloquently extols the importance of place). Jokes about environmental activists who travel the world to tell people to stay home miss the point. Traveling, with eyes open and mind engaged will broaden knowledge and attitudes, but a long-term commitment to home fosters a depth of knowledge and caring that is lost to a transient society.

One of the most striking memories from these travels was the Mountaintop Removal coal story. We both had a visceral reaction to seeing the ripped apart mountain beside Larry Gibson's place. That was a powerful moment. We could not appreciate the tragedy of Mountaintop Removal until we saw that hole at our feet and the walls of debris choking the stream channels. We cannot believe that anybody thinks it is a good idea. For people living there, like Larry, the loss is impossible to imagine.

For over 20 years Larry beat the drum about MTR, and did not give up. He still has hope. He did not just throw up his hands, but fought back. It became his life's work. He is knowledgeable and an eloquent speaker, who comes across so well, partly because he is simply a common man.

An overarching theme touched by coal in West Virginia, river remodeling in Korea, groundwater pumping in Kansas, and dam building in China, Turkey, Portugal, and California, was the mining mentality that focuses on resource extraction ahead of long-term sustainability. Wherever things had been damaged, it was inspiring to see renewal being sought and achieved. Sado Island, in Japan, best illustrated that theme, where an entire community works with a common goal to create a harmonious environment for humans and nature, one that will nurture the Japanese Crested Ibis.

Our journeys began at Mono Lake, another place that became a model of successful recovery against seemingly hopeless odds. While not all the stories on the 38th Parallel have happy endings, at least, not yet, we found many reasons for hope. Dedicated people were not willing to simply accept the unacceptable, even though their challenges seemed daunting.

We found a world of bountiful water environments that justify all that dedication and persistence.

Acknowledgments

Nick and Ryan Carle, Albert and Doris Broughton, for help with logistics, house- and car-sitting, and for not worrying too much when we disappeared over the horizon, and Krista and Philip Devaul, our hosts in Washington D.C.; Sally Gaines and Rick Kattelmann for reading the draft manuscript.

Rasa Gustaitis, Editor of *California Coast & Ocean;* a condensed version of the California trek chapter appeared in that journal in Autumn, 2009.

In order of appearance in the book or during our travels:

Korea: KFEM, Chooney Kim, Yong-un Ma, Hang Jin Lee, Jong-Hak Park, Naree Jeong; www.krem.or.kr
> Allen Lampson, Jane Hallinger, Pasadena Sister Cities Committee; www.passcc.org; Peace Hospital Foundation, Dr. Jung Nam
> Jung Rok An, DMZ Ecological Inst., www.ecodmz.or.kr
> Jang Moon Young "Moon", Paju City coordinator and Seung-Pyo Hong, Vice-Mayor of Paju City; www.paju.go.kr
> DMZ Forum: www.dmzforum.org
> Se-Yung Jun and Ms. Min-Hwa, U.C. Press representatives, Information and Culture Korea, www.ick.co.kr; Ms.Hae-Young Shon; Yoo and Kyobo Book Center import manager

China: Kinder (Jinde) Shu, EDG Earth Expedition (Beijing); www.china-edg.com
> Abdul Wayit; Silk Road Tour Guide; www.kashgartours.com
> Deputy Director Li Juanzhang and Shen Kai, engineer, at the Yellow River Delta (*Huanghe Sanjiaozhou*) National Nature Reserve
> Ran Liping, Green Camel Bell; www.gcbcn.org/en, introduced to us by Xiu Min Li, Pacific Environment; Zhang Songlin, Northwest Normal University, Geography and Environmental Sciences
> Fafu Li, Qinghai Institute of Salt Lakes; http://english.isl.cas.cn/au/bi/200905/t20090522 _3249.html; introduced to us by Dr. Robert Jellison, Sierra Nevada Aquatic Research Laboratory; http://vesr.ucnrs.org /pages/snarlmoreinfo.html
> Liu Qiang, Vice-general Director, China Association of National Parks and Scenic Sites, www.npachina.com
> Jim Harris and George Archibald, International Crane Foundation; www.savingcranes.org
> Dr. Rick Kattelmann, for advice and companionship in western China
> John and Nancy Walter of Mammoth Lakes, California, and Karen and Paul Amstutz, of Yosemite, intrepid travelers in Asia

Turkmenistan: Berkeli Atayev, Chief of International Relations, Marketing and Advertising Department; State Committee of Turkmenistan for Tourism and Sport www.tourism-sport.

gov.tm; introduced to us by Patrick and Lynn de Freitas, Friends of Great Salt Lake; www.fogsl.org

Turkey: Ipek Tasli, Initiative to keep Hasankeyf Alive; www.hasankeyfgirisimi.com; Thomas Wenidoppler, Stop Ilisu Campaign; http://m-h- s.org/ilisu

Mustafa Kamal Caddesi, Nemrut Tours from Malatya; www.nemruttours.com

Omar Digirmenci, "Mr. Happy," proprietor of the Liman Hotel, Kusadasi; www.limanhotel.com

Greece: Margaret and John Chaconas, who welcomed us to their house near Athens

Paul Tsaros, Archelon Sea Turtle Rescue Center

Sicily: Anna Giordano, Mediterranean Association for Nature, www.migrazione.it

Spain: Ranger Antonio Saez, at the Torrevieja and La Mata lagoons

Javi and Virginia Grijalbo; http://javiergrijalbo.blogspot.com, our naturalist guides across southern Spain

German Garrote Alonso, Antonio Rivas, Astrid Vargas, Miguel Angel Simon; www.lynxexsitu.es; www.lifelince.org

Portugal: José Martins, Quercus - Associação Nacional de Conservação da Natureza; www.quercus.pt

United States

Chesapeake Bay: Joelle Buffa, biologist at the Chincoteague National Wildlife Refuge, and her husband, Clyde Morris, for educating us and sharing their home; introduced to us by Emilie Strauss; Myrna Cherrix; Assateague lighthouse historian, www.fws.gov/northeast/chinco

Ann Gilmore, Coastal Virginia Wildlife Observatory, www.cvwo.org

Jack Russell and Krispen Parke, environmental educators with the Chesapeake Bay Foundation; www.cbf.org

John Tippett, Friends of Rappahanock, www.riverfriends.org

Jim Long and Bonnie Bick, Mattawoman Watershed Society, http://mattawomanwatershedsociety.org/

West Virginia:

Justin Reynolds, guide, and Skip Heater, owner, New & Gauley River Adventure; www.gauley.com

Larry Gibson, Keeper of the Mountain Foundation, http://mountainkeeper.blogspot.com/

Roland Micklem and Andrew Munn, Seniors March to end Mountaintop Removal

Kentucky:

Malcolm and Judy Dalgleish, who hosted us in Indiana and introduced us to Wendell and Tanya Berry

Akiko Mcvarish Gothard, Heatherway, equine management; Bev and John Passarello, Passarelo Thoroughbreds

Larry Meadows, Red River Museum

Missouri:

Terry Cunningham, Jason Green, Brandon Kuhn of the
Pioneer Forest; www.pioneerforest.org

Kansas:

Eric Patternson, Tallgrass Prairie National Preserve;
www.nps.gov/tapr

Rachel Laubhan and Ranger Barry Jones, Quivira National
Wildlife Refuge; www.fws.gov/quivira

Mark Rude and Jason Norquist, Southwest Kansas
Groundwater Management District No. 3; www.gmd3.org

Nancy Harness, Mayor of Garden City; www.garden-city.org,
introduced to us by Molly Desbaillets

Mark Sexson, Kansas Dept of Wildlife & Parks;
www.kdwp.state.ks.us

Tom Norman, Sandsage Bison Range; www.kdwp.state.ks.us/
newsKDWP-Info/Locations/Wildlife-Areas/Region-3/
Sandsage-Bison-Range

Colorado:

Shannon Nelson and Brett Pyle, our hosts on a snowy night in
Salida

Vicki Klein, Betsy Steinberg, John and Ann Graham, Chafee
County Citizens for Sustainability,
http://ccfsustainability.org

Paul Robertson, The Nature Conservancy; www.nature.org

Walter Wright, Telluride 350, www.climategroundzero.org

Utah:

Elizabeth "Liz" Glowiack and Ken Pill, Moab TAC Team and
Donald Metzler, U.S. Department of Energy,
www.gjem.energy.gov/moab

Kimberly Schappert, Up the Creek Campground, Moab Trails
Alliance

Nevada:

Glennon Zelch, Pioche

California:

Geoff McQuilkin, Mono Lake Committee; www.monolake.org

Dr. Connie Millar, US Forest Service, Pacific Southwest Research
Station; www.fs.fed.us/psw

Spreck Rosekrans, Restore Hetch Hetchy; www.hetchhetchy.org

Kurt Stegen, Friends of the River; www.friendsoftheriver.org

John Knott, David Martin, Marty Knott and Skippy the
dog, sailboat captains and crew

John Cain, Natural Heritage Institute; www.n-h-i.org

Marc Holmes, The Bay Institute; www.bay.org

Stan and Sharon Bluhm, Mono Lake volunteers who opened their
house to us in San Anselmo, and Richard and Brenda
Nichols, Coastwalk California; www.coastwalk.org

Don Neubacher, Point Reyes National Seashore; www.nps.gov/pore

Allen Fish, Golden Gate Raptor Observatory; www.ggro.org

David Wimpfheimer, www.calnaturalist.com; guide for Marin
Audubon Society trip to Farallones;
http://marinaudubon.aviandesign.net

257

Japan:
Tsunao Watanabe, Director General, Nature Conservation Bureau, Ministry of the Environment; Tetsuro Uesugi, Director, National Park Division; Shinjiro Sasaki, Keisuke Takahashi, and Kohei Sasabuchi, Assistant Directors National Park Division; Hiroshi Horiuchi, Wildlife Division; Kohei Sasabuchi

Dr. Tsuneo Sekijima, Dept. of Environmental Science, Niigata University; Dr. Chihiro Endo, Toki & Nature Restoration Research Center, Sado Island; Taeko Ishima, graduate student at Niigata University

Dr. Yoichi Kawaguchi, Tokushima University

Kei Osada, Ranger, Crested Ibis Conservancy, Sado Island; www.visitsado.com/en/00sp/0809/toki1.shtml

Akira Wada, Asian River Restoration Network; www.a-rr.net

Bibliography

Ashworth, William. 2006. *Ogallala blue: water and life on the high plains.* NY: W.W. Norton.

Audubon, John James. 1831. *Ornithological Biography/Vol1/The Ohio.* Edinburgh: Adam Black. http://books. google.com/booksid=cuRAAAAAcAAJ&source=gbs_navlinks. (accessed February 23, 2011).

Baker, Aryn. 2015, September 7. "How Climate Change is Behind the surge of Migrants to Europe." *Time. http://time.com/4024210/climate-change-migrants/.*

Baumer, Christoph. 2003. *Southern silk road: in the footsteps of Sir Aurel Stein and Sven Hedin.* Bangkok: Orchid Press.

Berry, Wendell. 1991. *The unforeseen wilderness: Kentucky's Red River Gorge.* Emeryville, CA: Shoemaker Hoard.

_____. 1998. *The selected poems of Wendell Berry.* Berkeley: Counterpoint.

_____. 2002. *The art of the commonplace.* Berkeley: Counterpoint.

Bonavia, Judy. 2008 [1998]. *The silk road; Xian to Kashgar.* Hong Kong: Odyssey Publications, Ltd.

Brown, Lester R. 2009. *Plan B 4.0, Mobilizing to Save Civilization.* New York: W.W. Norton & Company.

Buck, Pearl S. 1963 [1931]. *The good earth.* NY: Cardinal/Pocket Books.

Burdick, Charles F. 2003. *Stories of river pilots on the great western rivers.* Grand Tower, IL: BurPublishing.

California Department of Water Resources. 2006. "Hetch Hetchy Restoration Study." http://hetchhetchy.water.ca.gov/ docs/Hetch_Hetchy_Restoration_Study_Report.pdf (Accessed December 6, 2010).

Carle, David. 2016. *Water and the California dream.* Berkeley: Counterpoint Press.

_____. 2015 [2004]. *Introduction to water in California.* Berkeley: University of California Press.

Carle, David with Janet Carle. 2009. "Exploring California's 'Water Line.'" *California Coast & Ocean.* Vol. 25. No. 2: (11-20).

Chamberlin, J. Edward. 2006. *Horse: how the horse has shaped civilizations.* NY: BlueBridge.

Cherniack, Martin. 1986. *The Hawk's Nest incident: America's worst industrial disaster.* New Haven, CT: Yale University Press.

Clarke, Thurston. 1988. *Equator: a journey.* NY: Avon Books.

Dai Qing. 1998. *The river dragon has come! The Three Gorges Dam and the fate of China's Yangtze River and its People.* International Rivers Network. Armonk, NY: M.E. Sharpe.

Economy, Elizabeth C. 2010 [2004]. *The river runs black: the environmental challenge to China's future.* Ithaca: Cornell University Press.

Egleston, N.H. 1909. "Granting use of Hetch Hetchy to City of San Francisco." Report to 60th Congress. February 8, 1909. Washington, D.C.

Fagan, Brian. 2011. *Elixer, A History of Water and Humankind.* New York: Bloomsbury Press.

Fatakuchi, Kazuko. 2010. Transforming agriculture and economy to save the Japanese Crested Ibis: Sado Island." JFS Newsletter, Japan Ministry of the Environment. No. 97 (September 2010). www.japanfs.org/en/mailmagazine/newsletter/pages/030310.html (accessed March 4, 2011).

Geredieu, Ross. 2009. "Extent of Mountaintop Mining in Appalachia as of 2008." Boone, NC: Appalachian Voices. www.ilovemountains.org/reclamation-fail/mining-extent-2009/MTR_Peaks_Update.pdf. Accessed 2/20/2010

Gow, A.S. F. and Page, DL (editors). 1968. *The Greek anthology: the garland of Philip and some contemporary epigrams.* London: Cambridge University Press.

Heat-Moon, William Least. 1999. *PrairyErth (a deep map): an epic history of the tallgrass prairie country.* Boston: Mariner Books.

Hessler, Peter. 2002. *River town: two years on the Yangtze.* NY: Perennial (Harper Collins).

Horton, Tom. 1996. *An island out of time: a memoir of Smith Island in the Chesapeake.* NY: W.W. Norton & Co.

_____. 1991. *Turning the tide: saving the Chesapeake Bay.* Washington D.C.: Chesapeake Bay Foundation and Island Press.

Irving, Washington. 1994. *Tales of the Alhambra.* Ediciones Miguel Sanchez. Granada, Spain.

Ivanova, Nadya. 2011. "Along China's Yellow River, Trades Tie Water Savings On Farms to Industrial Expansion" Circle of Blue. www.circleofblue.org/waternews/2011/world/from-agriculture-to-industry-efficiency-upgrades-transfer-water-use-rights-on-china. (accessed March 9, 2011).

Kreutzmann, Hermann. 2003. "Ethnic minorities and marginality in the Pamirian Knot: survival of Wakhi and Kirghiz in a harsh environment and global contexts." *The Geographical Journal.* Vol. 169, No. 3. September. (215–235).

Latham, Ronald (translator). 1982 [1958]. *The travels of Marco Polo.* NY: Abaris Books.

Los Angeles *Times.* 1974. "It's a Hoax, All right." Editorial. October 25; Part 2 (6).

Malloy, Daniel. January 14, 2011. "EPA Vetoes Mining Permit in West Virginia." *Pittsburgh Post-Gazette.* www.post-gazette.com/pg/11014/1117862-113.stm. (accessed February 22, 2010).

Matthiessen, Peter. 2001. *The birds of Heaven: travels with cranes.* NY: North Point Press.

Mertha, Andrew C. 2008. *China's Water Warriors: Citizen Action and Policy Change.* Ithaca: Cornell University Press.

Mount, Jeffrey F. 1995. *California rivers and streams.* Berkeley: University of California Press.

Muir, John. 1912. *The Yosemite.* New York: Century Company.

Nakamura, Keigo, Klement Tockner, and Kunihiko Amano. 2006. "River and Wetland Restoration: Lessons from Japan." BioScience, Vol. 56 No. 5. May (419-429).

Nash, Ogden. 1931. *Free wheeling.* NY: Simon and Schuster.

Nation, Allan H. 2010. "Allan's Observations." *Stockman Grass Farmer.* 10(1).

Opie, John. 2000. *Ogallala: water for a dry land.* Lincoln: University of Nebraska Press.

Pacific Institute, "Bottled Water and Energy."www.pacinst.org/topics/ water_and_sustainability/bottled_water/bottled_water_and_ energy.html. (accessed February 25, 2010).

Palmer, Tim. 1982. *Stanislaus: the struggle for a river.* Berkeley: University of California Press.

Paul, Kathleen. 2011. "The Story Behind the Fund." Matsushima Relief Fund. https://sites.google.com/site/matsushimafund/ background. (Accessed September 19, 2011)

Polo, Marco. 982 (1958). *The Travels of Marco Polo: translated and introduced by Ronald Latham.* New York: Abaris Books, Inc.

Powell, John Wesley. 1879. "Report on the lands of the arid region of the United States." Washington, D.C.: Government Printing Office.

Prosek, James. 2003. *Fly-Fishing the 41st: around the world on the 41st parallel.* NY: HarperCollins.

Ramzay, Austin. 2009. "Heroes of the Environment 2009: Zhao Zhong." *Time.* September 22.

Righter, Robert W. 2005. *The battle over Hetch Hetchy: America's most controversial dam and the birth of modern environmentalism.* Oxford: Oxford University Press.

Rosenburg, Matt. T. "Largest Cities Through History." 2011. About.com Geography. http://geography.about.com/library/ weekly/aa011201a.htm. (accessed October 8, 2011).

Schneider, Keith. 2011. "Choke Point China: Confronting Water Scarcity." Circle of Blue. February 15, 2011. www.circleofblue.org /waternews/2011/world/choke-point- china—confronting-water-scarcity-and-energy-demand-in-the-world's-largest-country/ (accessed February 22, 2011).

Sekijima, Tsuneo, Yoichi Kawaguchi, Tadashi Miyashita, Yasuhiro Mitani, and Ryugo Watanabe. 2010. *Messages from Japan's Green Pioneers: Living in Harmony with Nature.* Translated into English from *Environmental Research Quarterly* by the Hitachi Environment Foundation. Tokyo: Ministry of the Environment, Government of Japan.

Shea, Kelly. 2011. "Circle of Blue Infographic: Going the Distance, From Ashgabat to Whyalla—10 Cities Pumping Water From Afar." Circle of Blue. February 1, 2011. www.circleofblue.org/ waternews/2011/world/infographic-going-the-distance/ (accessed February 3, 2011).

Simonds, Wm. Joe. 1994. "New Melones Unit Project History." U.S. Bureau of Reclamation. www.usbr.gov/projects/Project.jsp?proj_ Name=New%20Melones%20U%20Project&pageType=Project HistoryPage (accessed February 29, 2010).

Smith, Andrew T. and J. Marc Foggin. 1999. "The plateau pika (*Ochotona curzoniae*) is a keystone species for biodiversity on the Tibetan plateau." *Animal Conservation.* 2: 235-240.

Spangler, Patricia. 2008. *The Hawks Nest tunnel: an unabridged history.* Proctorville, OH: Wythe-North Publishing.

Takano, Koichiro. 2010. "Declaring 'Biological Survey Days." In "Sado City's Approach to Economic and Environmental Synergies." Sado City, Niigata Prefecture, Japan.

Tiller & Toiler. 2009. "Fort Larned 150th Anniversary 1859-2009." Special Commemorative Section in Tiller & Toiler. October 11. Larned, Kansas.

Twain, Mark. 1873. *Roughing It.* Hartford, CN: American Publishing Co.

U.S. Environmental Protection Agency. 2009. "The Effects of Mountaintop Mines and Valley Fills Aquatic Ecosystems of the Central Appalachian Coalfields." DRAFT. Office of Research and Development, National Center for Environmental Assessment, Washington, DC. EPA/600/R-09/138A.

U.S. Geological Survey. 2011. *Poster of the Great Tohoku Earthquake (northeast Honshu, Japan) of March 11, 2011 - Magnitude 9.0.* http://earthquake.usgs.gov/earthquakes/eqarchives/poster /2011/20110311.php (accessed September 27, 2011).

U.S. National Park Service. 1988. "Alternatives for Restoration of Hetch Hetchy Valley Following Removal of the Dam and the Reservoir." www.sierraclub.org/ca/hetchhetchy/ nps_hh_restoration.pdf (accessed December 10, 2010)

Vervaeck, Armand and James Daniell. 2011. "CATDAT Sendai Japan Total Situation Report v37." Earthquake Report.com. http://earthquake-report.com/2011/08/04/japan-tsunami- following-up- the-aftermath-part-16-june/ (accessed October 1, 2011)

Ward Jr, Ken. 2009. "Byrd blasts Massey 'arrogance' over school." *Charleston Gazette.* Oct. 8, 2009: 1.

Xinhua News Agency. 2008. "Massive program launched to save Qinghai Lake." May 26, 2008. www.chinadaily.com.cn/ china/2008-05/26/ (accessed February 6, 2010).

Yu, Junqing and Lisa Zhang. 2008. *Lake Qinghai: Paleoenvironment and Paleoclimate.* Beijing: Science Press.

Index

Index

Index

Index

Index

CPSIA information can be obtained
at www.ICGtesting.com
Printed in the USA
LVHW021941070520
655176LV00005B/45